By Brian Hare and Vanessa Woods
The Genius of Dogs
Survival of the Friendliest

By Vanessa Woods
Bonobo Handshake

Edited by Brian Hare (with Shinya Yamamoto)
Bonobos: Unique in Mind, Brain, and Behavior

Survival
of the
Friendliest

Survival of the Friendliest

Understanding Our Origins
and Rediscovering
Our Common Humanity

Brian Hare
and
Vanessa Woods

RANDOM HOUSE · NEW YORK

Published in the United States by Random House, an imprint and
division of Penguin Random House LLC, New York.

RANDOM HOUSE and the HOUSE colophon are registered trademarks of
Penguin Random House LLC.

LIBRARY OF CONGRESS CATALOGING-IN-PUBLICATION DATA
Names: Hare, Brian, author. | Woods, Vanessa, author.
Title: Survival of the friendliest / Brian Hare and Vanessa Woods.
Description: First edition. | New York: Random House, [2020] | Includes
bibliographical references and index. |
Identifiers: LCCN 2019049396 (print) | LCCN 2019049397 (ebook) |
ISBN 9780399590665 (hardcover) | ISBN 9780399590672 (ebook)
Subjects: LCSH: Evolutionary psychology. | Social evolution. | Human
evolution. | Friendship. | Empathy.
Classification: LCC BF698.95 H37 2020 (print) |
LCC BF698.95 (ebook) | DDC 155.7—dc23
LC record available at https://lccn.loc.gov/2019049396
LC ebook record available at https://lccn.loc.gov/2019049397

Printed in the United States of America on acid-free paper

randomhousebooks.com

246897531

First Edition

Book design by Susan Turner

For all humans

E pluribus unum

Contents

Introduction   xiii

1   Thinking About Thinking   3

2   The Power of Friendliness   18

3   Our Long-Lost Cousins   38

4   Domesticated Minds   55

5   Forever Young   78

6   Not Quite Human   102

7   The Uncanny Valley   122

8   The Highest Freedom   150

9   Circle of Friends   186

Acknowledgments   199

Notes   203

Image Credits   253

Index   255

Introduction

It was 1971, seventeen years after *Brown v. Board of Education* declared that segregation in schools was unconstitutional, and schools all over the country were still in turmoil.

Minority children often had to catch buses from one side of town to another, which meant they had to get up two hours earlier than white children. If they could afford to, white families sent their children to private schools, leaving only the poorest white children in the public school system. In the classroom, there was so much hostility between racial groups that children had very little energy for learning. Educators, parents, policy makers, civil rights activists, and social workers watched in dismay.

Carlos* was in the fifth grade at a public school in Austin, Texas. English was his second language. He answered questions with a stammer, and when the other children mocked him, he stammered even more. He became withdrawn, rarely speaking at all.

* Not his real name

Many social scientists had predicted that school desegregation would be an unqualified success. It was assumed that once all children were on an equal footing in the classroom, white children would leave school less racist, not just toward people of color in their schools, but toward those they encountered throughout their lives. Minority children would receive a first-class education, which would set them up for successful careers.

However, when the psychologist Elliot Aronson looked in on Carlos and his classmates, he spotted a fundamental problem. The children in that classroom were not on an equal footing. The white children were better prepared, better equipped, and better rested. Many teachers were teaching minority children for the first time and were as bewildered by their new charges as their white students were. Carlos's teacher, seeing how badly he was being teased and not wanting to put him on the spot, no longer called on him, inadvertently isolating him further. Other teachers did not want minority children in their classroom. If they did nothing to encourage the white children's merciless taunting, neither did they do anything to stop it.

In a traditional classroom structure, children are in constant competition for their teacher's approval. This inherent conflict—in which the success of one child threatens the success of another—can foster a toxic environment, and integration exacerbated this issue. Many of the white children had been at their school for years. They saw minority children as invaders, and inferior ones at that. The minority children understandably felt threatened by their hostility.

Aronson convinced Carlos's teacher to try something

new. Instead of the teacher holding court over the class, asking questions, singling out some students and neglecting others, Aronson suggested transferring a small portion of knowledge, and its associated power, to each student.

Carlos's class was studying the journalist Joseph Pulitzer. Aronson broke the class into groups of six. Each member of Carlos's group had to learn about one stage of Pulitzer's story, and at the end of the exercise, the children were tested on his whole life. Carlos was in charge of Pulitzer's middle years. When it was his turn to say what he had learned, he stammered as usual, and the other children made fun of him. Aronson's assistant casually remarked, "You can say things like that if you want to, but it's not going to help you learn about Joseph Pulitzer's middle years, and you'll have an exam on Pulitzer's life in about twenty minutes."

The children quickly realized that they weren't competing against Carlos, they needed him. Making Carlos nervous made it harder for him to explain what he'd learned, so they became sympathetic interviewers, carefully drawing out what he knew. After several weeks of working like this on various projects, Carlos became more comfortable around the other children, and they grew friendlier.

Aaronson had introduced what's known as the "jigsaw" method, in which each child in a group has a piece of knowledge to contribute to a coherent lesson.[1] Working this way for just a few hours each week had powerful effects. After only six weeks, Aronson found that both the white and minority children liked the members of their jigsaw group, regardless of their race, more than they liked the other children in their classrooms. They liked school better, and their self-esteem improved. Jigsaw children empathized with others more easily and academically outperformed children in competitive classrooms. Minority children showed the biggest improvements of all. Cooperative learning methods were repeated with similar results, time after time, in hundreds of different studies and in thousands of classrooms around the United States.[2, 3, 4, 5]

SURVIVAL OF THE FITTEST

Cooperation is the key to our survival as a species because it increases our evolutionary fitness. But somewhere along the way, "fitness" became synonymous with physical fitness. In the wild, the logic goes, the bigger you are, and the more willing you are to fight, the less others will mess with you and the more successful you will become. You can monopolize the best food, find the most attractive mates, and have the most babies. Arguably, no folk theory of human nature has done more harm—or is more mistaken—than the "survival of the fittest." Over the past century and a half, it has been the basis for social movements, corporate restructuring, and extreme views of the free market. It has been used to argue for the abolition of government, and to judge groups of people as inferior, and then justify the cruelty that results. But to Darwin and modern biologists, "survival of the fittest" refers to something very specific—the ability to survive and leave behind viable offspring. It is not meant to go beyond that.

The idea that the strong and ruthless will survive while the weak perish became cemented in the collective consciousness around the publication of the fifth edition of Charles Darwin's *Origin of Species* in 1869, in which he wrote that as a proxy for the term "natural selection," "Survival of the Fittest is more accurate, and is sometimes equally convenient."

Darwin was constantly impressed with the kindness and cooperation he observed in nature, and he wrote that "those communities, which included the greatest number of the most sympathetic members, would flourish best and rear the greatest number of offspring."[6] He and many of the biologists who followed him have documented that the ideal way to win at the evolutionary game is to maximize friendliness so that cooperation flourishes.[7, 8, 9]

The idea of "survival of the fittest" as it exists in the popular imagination can make for a terrible survival strategy. Research shows that being the biggest, strongest, and meanest animal can set you up for a lifetime of stress.[10] Social stress saps your body's energy budget, leaving a weakened immune system and fewer offspring.[11] Aggression is also costly because fighting increases the chance that you will be hurt or even killed.[12, 13] This kind of fitness can lead to alpha status, but it can also make your life "nasty, brutish and short."[14, 15, 16] Friendliness, roughly defined as some kind of intentional or unintentional cooperation, or positive behavior toward others, is so common in nature because it is so powerful. In people, it can be as simple as approaching someone and wanting to socially interact or as complicated as reading someone else's mind in order to cooperatively accomplish a mutual goal.[7]

It is an ancient strategy. Millions of years ago, mitochondria were free-floating bacteria until they entered larger cells. Mitochondria and the larger cells joined forces and became the batteries that power cell function in animal bodies.[17] Your microbiome, which, among other things, allows your body to digest food, make vitamins, and develop organs, is the result of similar, mutually beneficial partnerships between microbes and your body.[18] Flowers appeared later than most plants, but their mutually beneficial relationship with pollinating insects made them so successful that they now dominate the landscape.[19] Ants, estimated to have the same mass as a fifth of all the other terrestrial animals on earth combined, can form superorganisms of up to 50 million individuals that function as a single social unit.[20]

Each year I* challenge my students to use evolutionary theory to solve the world's problems. In this book, we gave ourselves the same challenge. This is a book about friendliness and how it came to be an advantageous evolutionary strategy. It is a book about understanding animals—and here dogs play the starring role—because in doing so, we can better understand ourselves. It is also an exploration of the flip side of our friendliness: the capacity to be cruel to those who aren't our friends. If we can develop an understanding of how this dual nature evolved, we can find powerful new ways to address the social and political polarization that endangers liberal democracies around the world.

THE FRIENDLIEST HUMAN

We tend to think of evolution as a creation story. Something that happened once, long ago, and continued in a linear fashion. But evolution is not a neat line of life forms progressing toward the "perfection" of *Homo sapiens*. Many species have

---

* Although Brian and Vanessa contributed equally to this book, Brian's research is the main focus, so we will use "I" to refer to Brian throughout.

been more successful than we are. They have lived for millions of years longer than we have and spawned dozens of other species still alive today.

The evolution of our own lineage, since we split from our common ancestor with bonobos and chimpanzees around 6 to 9 million years ago, produced dozens of different species within our genus, *Homo*. There is fossil and DNA evidence that for most of the approximately 200,000 to 300,000 years that *Homo sapiens* has existed, we shared the planet with at least four other human species.[21] Some of these humans had brains that were as big as, or bigger than, our own. If brain size was the main requirement for success, these other humans should have been able to survive and flourish as we did. Instead, their populations were relatively sparse, their technology—though impressive compared to that of nonhumans—remained limited, and at some point, all of them went extinct.

Even if we had been the only humans with big brains, we would still have to explain the gap of at least 150,000 years between our appearance in the fossil record and the explosion in our population and culture. Though the physical features that distinguished us from other humans marked us early in our evolution, we were still culturally immature for at least 100,000 years after we appeared in Africa. There were tantalizing glimpses of the technology we would become famous for: blades carefully worked to symmetrical points, objects painted with red pigment, bone and shell pendants. For thousands of years, these innovations flickered but did not take hold.[22, 23, 24]

If, a hundred thousand years ago, you were setting the

odds for which human species would be the last one standing, we would not have been a clear winner. A more likely contender would have been *Homo erectus,* who left Africa as early as 1.8 million years ago to become the most widespread species on earth. *Homo erectus* were explorers, survivors, and warriors. They colonized most of the planet, and somewhere along the way they learned to control fire, which they used for warmth, protection, and cooking.

Homo erectus were the first humans to skillfully use advanced stone tools, including the Achulean hand ax made from raw materials like quartz, granite, and basalt.[25, 26] The type of rock dictated its sculpting method—chipping or flaking—and the results were tear-shaped, razor-sharp tools so beautifully crafted that when people found them thousands of years later, they thought the stones had supernatural powers. *Homo erectus* saw the rise and fall of many other human species, and they existed longer than any other humans, including us.

At one hundred thousand years ago, we were still using the same hand ax *Homo erectus* had invented a million and a half years before we emerged. Genetic evidence suggests that our population might have diminished to a point near extinction.[27, 28, 29] *Homo erectus* probably thought we were just another short-lived novelty of the Pleistocene.

Fast-forward to 75,000 years ago. *Homo erectus* were still around, but their technology had not advanced much, and you might guess the Neanderthals would succeed instead. Neanderthals had brains that were as big as or bigger than ours. They were as tall as we were, but heavier, and most of their extra weight was muscle. Neanderthals ruled the Ice

Age. Although technically omnivores, they tended to be carnivorous, which means they had to be skillful hunters. Their main weapon was a long, heavy spear meant to be thrust at close range. Carnivores usually hunt animals smaller than themselves. Neanderthals hunted every large herbivore in the Ice Age, mostly red deer, reindeer, horses, and bovines, and the occasional mammoth, all of which were considerably more powerful than humans.[23]

Neanderthals were far from grunting cavemen. We share with them a variant of the FOXP2 gene thought to be responsible for the fine motor movements needed for speech.[30] They buried their dead, cared for their sick and injured, painted themselves with pigment, and adorned themselves with jewelry made of shells, feathers, and bone. One Neanderthal man was found buried with almost three thousand pearls decorating his animal hide clothing that had been expertly stretched and stitched together.[31] They made cave paintings depicting mythical creatures. Toward the end of their time, they had many of the same tools we did.[23]

When *Homo sapiens* first met Neanderthals, the Neanderthal population was the largest it had ever been. Because they were cold adapted, they replaced us as we fled Europe in the face of oncoming glaciers. If 75,000 years ago you were going to put money on who would survive the uncertain climate of the future millennia, Neanderthals would have been a good bet.

However, by 50,000 years ago, the tide was turning in our favor. While the Achulean hand ax had served every human species for more than a million years, we had developed a far more complex toolbox. Improving upon the wooden thrusting spears of Neanderthals, we developed projectile weapons like the spear thrower—a two-foot wooden handle that launched six-foot arrow-like shafts. The arrows, usually tipped with sharpened stone or bone, were hollowed out at one end and set onto the wooden shaft.[32] The physics worked on the same principle as a Chuckit—the dog owners' ball thrower. Even if you were very strong, you would be able to throw a standard spear only a few feet by hand. The stored energy in the shaft of a spear thrower could propel the shaft more than 300 feet at better than 100 miles per hour. Spear throwers revolutionized hunting, allowing us to graduate beyond human-sized herbivores and hunt prey that flew, swam, and climbed trees. We could kill mammoths without the risk of being stamped by their feet or speared by their tusks. Spear throwers also revolutionized the way we protected ourselves. We could launch a spear at a saber-toothed cat or hostile human and cause serious injury from a safe distance. We made sharp points for weapons, tools for engraving, blades for cutting, and drill bits for piercing. We had boned harpoons; nets for fishing and traps; and snares for birds and small mammals. Neanderthals, for all their hunting prowess, were never more than mid-ranking carnivores. With our new technology, we became the ultimate predator, largely immune to predation by other species.

We ventured out of Africa and spread rapidly across Eurasia. We may even have reached as far as Australia within a few thousand years. This arduous crossing would have required planning and packing food for an indefinite journey, taking tools that could repair unforeseen damage and catch unfamiliar food, and solving future imagined problems like replenishing drinking water at sea. These early sailors had to be able to communicate in detail, leading some anthropologists to hypothesize that by this time, we already had full-blown language.[33]

Most remarkably, these sailors had to infer that there was something beyond the horizon. Perhaps they had studied the patterns of migratory birds or seen the smoke of natural bushfires in the distance. Even if this was the case, they would have had to imagine that there was somewhere to go.

By 25,000 years ago, the odds were clearly with us. Instead of operating as nomadic wanderers, we lived in more permanent camps filled with hundreds of people. Camps were organized by function: Discrete areas were allotted for butchering, cooking, sleeping, and dumping garbage. We were well fed and had grinding and pounding tools that allowed us to process and treat food that would otherwise have been inedible or even toxic. We had fire pits for cooking, ovens for baking, and ways to store food in lean times.[33]

Instead of draped or loosely tied fur, we had real clothing made possible by fine bone needles. Snugly fitting snowsuits meant we were better able to withstand the cold without evolving calorie-hungry bodies like the Neanderthals.[34] Thus equipped, we could push northward even in freezing glacial periods and eventually head toward the Americas—the first humans ever to make the journey.

But this period of time, now known as the Upper Paleolithic, was remarkable for more than just an upgrade in weapons and living conditions.[35] It was around this time that we began to leave evidence of unique forms of cognition, especially our expanding social networks.[36] Jewelry made from shells has been found hundreds of miles inland, implying that an object with no practical value was either worth carrying some distance or was obtained from someone else who had traveled on one of our first trade routes.[37, 38]

We painted animals on rocks so skillfully that the contours of the stone rippled beneath their bodies and gave them a third dimension. In what can be regarded as the creation of protocinema, a cave wall bears the illustration of a bison with eight legs that would have seemed to gallop in firelight. We even seem to have illustrated with sound: horses' mouths open in a whinny, lions depicted in midroar, and rhinoceroses butting their heads ferociously enough that you can almost hear the clash of their horns. We not only imitated life, we imagined and portrayed mythical creatures—a woman with the head of a lion, a man with the body of a bison.[39, 40]

This was behavioral modernity: We both looked and acted like modern humans. Our culture and technology had suddenly become far more powerful and sophisticated than that of any other human. But how? What happened to us, and why did it happen only to us?

What allowed us to thrive while other humans went extinct was a kind of cognitive superpower: a particular type of friendliness called cooperative communication. We are experts at working together with other people, even strangers. We can communicate with someone we've never met about a shared goal and work together to accomplish it. As you would expect, chimpanzees are cognitively sophisticated in many of the ways humans are. But despite our many similarities, they struggle to understand when communication is intended to help them accomplish a shared goal. This means that as smart as chimpanzees are, they have little ability to synchronize their behavior, coordinate different roles, pass on their innovations, or even communicate beyond a few rudimentary requests. We develop all of these skills before we can walk or talk, and they are the gateway to a sophisticated social and cultural world. They allow us to plug our minds into the minds of others and inherit the knowledge of generations. They are the foundation for all forms of culture and learning, including sophisticated language, and it was dense groups of these cultured humans who invented superior technology. *Homo sapiens* were able to flourish where other smart human species didn't because we excel at a particular kind of collaboration.

When I began studying animals, I was so focused on social competition that it never occurred to me that communication or friendliness could be important for cognitive evolution, not just in animals but in ourselves. I thought increased skill in manipulation or deception could explain the evolutionary fitness of an animal. What I discovered is that being smarter is not enough. Our emotions play an oversized role in what we find rewarding, painful, attractive, or aversive. Our preferences for solving certain problems over others plays as important a role in shaping our cognition as our computing abilities. The most sophisticated social understanding, memory, or strategy will not facilitate innovation unless it is paired with the ability to communicate cooperatively with others.

This friendliness evolved through self-domestication.[7]

Domestication is not just a result of artificial selection

accomplished by humans choosing which animals to breed. It is also the result of natural selection. In this case, the selection pressure would be on friendliness—either toward a different species or toward your own. This is what we call self-domestication. Self-domestication gave us the friendly edge we needed to succeed as other humans went extinct. So far, we have seen this in ourselves, in dogs, and in our closest cousins, bonobos. This book is about the discovery that linked our three species together and helped us understand how we became who we are.

As humans* became friendlier, we were able to make the shift from living in small bands of ten to fifteen individuals like the Neanderthals to living in larger groups of a hundred or more. Even without larger brains, our larger, better-coordinated groups easily outcompeted other species of humans. Our sensitivity to others allowed us to cooperate and communicate in increasingly complex ways that put our cultural abilities on a new trajectory. We could innovate and share those innovations more rapidly than anyone else. Other humans species did not stand a chance.

But our friendliness has a dark side. When we feel that the group we love is threatened by a different social group, we are capable of unplugging the threatening group from our mental network—which allows us to dehumanize them. Where empathy and compassion would have been, there is nothing. Incapable of empathizing with threatening outsiders, we can't see them as fellow humans and become capable

---

* Authors' note: Unless otherwise noted, "human" refers to our species of human *Homo sapiens*.

of the worst forms of cruelty. We are both the most tolerant and the most merciless species on the planet.[7]

Dehumanizing rhetoric flourishes in the current United States Congress, which is more polarized now than it has been since the Civil War.[41] The former Republican representative Jim Leach, of Iowa, claimed that "In the Republican cloakroom, truly bizarre things are said about the Democrats."[42] The former Democratic senator Tom Daschle of South Dakota said that "these caucuses have become pep rallies . . . and it becomes a 'we,' 'they,' 'kill-'em' attitude."[42] Social media has made this brand of animosity public. When Donald Trump, Jr., was quoted as saying "A border wall is like a zoo fence protecting you from the animals," the Democratic congresswoman Ilhan Omar of Minnesota retorted, "The higher a monkey climbs, the more you see of its behind."

In the recent past, Washington was a friendlier place. President Ronald Reagan used to invite both Democrats and Republicans to the White House for drinks, "just to tell jokes."[43] Democrats and Republicans used to carpool from their hometowns to D.C., driving all night and taking turns at the wheel. "We'd argue like hell on the floor of the House of Representatives," said Dan Rostenkowski, a Democratic congressman from Illinois, "but we were out playing golf that night."[43] When Reagan called Speaker of the House Tip O'Neill after a particularly heated exchange, Tip said, "Old buddy, that's politics—after six o'clock we can be friends."[44]

That kind of Congress got things done. More bills were introduced and passed then than today. More people voted across the aisle. In 1967, Republicans and Democrats passed the Civil Rights Act, the most important social legislation in a century. Democrats worked with Republicans to pass Reagan's tax plan, the most significant tax reform in modern history.

Then, in 1995, a young Republican congressman from Georgia by the name of Newt Gingrich came up with a plan to loosen the hold Democrats had had on Congress for more than forty years. His theory was that as long as a Congress was working, people would not want to change the party in control of it. In his words, "You have to blow down the old order to establish a new order."[45]

One of Newt Gingrich's main tactics as Speaker of the House in the late nineties was to institute policies explicitly designed to make friendships between Republicans and Democrats difficult, if not impossible. He started by simply changing the Washington workweek from five to three days so Republican representatives would spend the majority of time in their home districts, connecting with constituents and fundraising. This move hindered cross-group friendships, since fewer congresspeople moved their families to Washington.[46] The political scientist Norman Ornstein wrote, "It used to be all the time that members were around on the weekends and they'd have dinner parties or they'd wind up with their kids at the same schools. . . . We just don't have that anymore."[42, 47]

On Capitol Hill, Gingrich forbade Republican cooperation with the Democrats, either in committees or on the floor of the House. When Republicans spoke about a Democrat or the Democratic Party, they were advised to use dehumanizing language that elicited disgust, to characterize the opposition with words such as "decay" and "sick."[48] Gingrich frequently compared the Democrats to the Nazis.[49] When he led the Republicans into this new and hostile territory, many Democrats eagerly followed suit. There were no more deals worked out behind closed doors, no bipartisan meetings or caucuses. Eventually the norms Gingrich introduced to the House took over the Senate's culture as well.[50]

As former senator Joe Biden of Delaware said of his relationship with the late Senator John McCain of Arizona, "John and I used to do debates in the nineties. We'd go over and sit with each other, literally sit next to each other on either the Democratic or Republican side of the floor . . . we . . . were chastised by the leadership of both our caucuses—why were we talking with each other and sitting with each other showing such friendship in the middle of debates . . . after the Gingrich Revolution in the nineties. They didn't want us sitting together, that's when things began to change."[51]

As comity dissolved on the hill, tools that had allowed negotiation and compromise were vilified. Pork barrel projects—projects funded by the federal government that benefit a relatively small number of people—fell out of fashion. Pork barrel projects might seem like a wasteful practice, but they are a tiny portion of the federal budget and have generally been crucial in pushing through vital legislation. The political scientist Sean Kelly found that the 2010 ban on pork barrel spending locked up the gears that make Congress turn.[52] After the ban was in place, nearly one hundred fewer laws were passed each year. Policy makers were less likely to succeed when they had no carrots to motivate compromise.

Like students in a competitive classroom, members of different parties saw how they no longer depended on one another, and they refused to cooperate.

Political rivals in a liberal democracy cannot afford to be enemies.[53] Socializing with your rivals humanizes them. Cooperation, negotiation, and trust—currently in short supply in Washington—become possible.

The self-domestication hypothesis is not just another creation story. It is a powerful tool that can help us short-circuit our tendency to dehumanize others. It is a warning and a reminder that in order to survive and thrive, we need to expand our definition of who belongs.

Survival
of the
Friendliest

Thinking About Thinking

When you were around nine months old, before you could walk or talk, you began to point. Of course, you could point soon after you were born, but at nine months, it started to mean something. It is a curious gesture. No other animal does it, even if they have hands.

Understanding the meaning of a point requires sophisticated mind reading. It generally means "If you look over there, you'll know what I mean."[1] But if I see you point to your head, there are many possible meanings. Are you referring to yourself? Are you saying I'm crazy? Did I forget my hat? A point can refer to something in the future or to something that used to be but is no longer.

Before you were nine months old, if your mom pointed, you likely looked at her finger. After nine months, you started to follow an imaginary line extending from her finger. By sixteen months, you would check that your mom was looking before you pointed, because you knew you needed her attention. By two years old, you knew what others saw and what they believed. You knew whether their actions were by accident or design. By age four, you could guess someone's thoughts so cleverly that for the first time, you could lie. You could also help someone if they had been deceived.[2]

Pointing is the gateway to reading other people's minds, to what psychologists call "theory of mind."[3] You will spend the rest of your life wondering what other people are thinking. The meaning of a hand brushed against yours in the dark. A raised eyebrow when you walk into a room. It will always be a theory, because you can never really know someone else's mind. Other people have the same abilities you do and can feint, fake, and lie.

Theory of mind allows us to engage in the most sophisticated cooperation and communication on the planet. It is crucial to almost every problem you will ever face. It allows you to time-travel and learn from people who lived hundreds and even thousands of years before you. Language is important but fairly useless if you do not know what your audience knows. You can teach only if you can remember what it is like not to know. The political party you vote for, the religion you follow, the sports you play, and every other experience that involves other people, living or dead, real or imagined, all rely on your theory of mind.

It is also the soul of your existence. Without it, love would be a cardboard cutout of itself, because what is love without the magic of knowing someone else feels the way you do? Theory of mind is the delight of moments when you both see something, then turn to each other and laugh. It is the comfort of finishing each other's sentences, and the peace in holding hands and saying nothing at all. Happiness is sweeter if you think the people you love are happy too. Grief is more bearable if you believe someone you lost would be proud of who you are.

Theory of mind is also the source of suffering. Hatred burns brighter if you are convinced someone intends you harm. Betrayal is more bitter when you can sift through a hundred memories for every subtle gesture that should have been a warning.

Every emotion we have enriches the lens through which we see the world. And though we "feel" these emotions in our chest, our gut, and the tips of our fingers, they live in our mind and are largely created from our theories about the minds of others.

DOG DAYS

My closest childhood friend was my dog Oreo. My parents gave him to me when I was eight years old, and he quickly grew from a puppy I could hold in my hands to a 70-pound Labrador with a wolfish appetite and a joy for life.

On warm nights, we would sit together on the front steps, his head on my lap. It never bothered me that he could not talk. I just enjoyed being with him, wondering what the world looked like through his eyes.

When I went to college at Emory, I discovered that exploring the animal mind was a serious scientific endeavor. I began working with Mike Tomasello, a psychologist who was an expert on theory of mind in children. Mike's experiments with babies connected their earliest theory of mind abilities with their ability to acquire all forms of culture—including language.[4]

Mike and I worked together for ten years, testing the theory of mind abilities of one of our two closest living relatives, chimpanzees. Before our experiments, there was no experimental evidence that any animal had theory of mind. But our research showed that the answer was more complicated.

Chimpanzees had some ability to map the minds of others. In our experiments, we found that not only did chimpanzees know what someone else saw, they knew what someone else knew, could guess what someone else might remember, and understood the goals and intentions of others. They even knew when someone else had been lied to.[2]

The fact that chimpanzees could do all these things put what they could *not* do into sharp resolution. Chimpanzees can cooperate. They can communicate. But they struggle to do both at the same time. Mike told me to hide a piece of food under one of two cups so that a chimpanzee would know that I had hidden the food, but not where. Then I would try to tell them which was the correct cup by pointing to it. Almost unbelievably, the chimpanzees, trial after trial, ignored my helpful gesture and could only guess. They became successful only after dozens of trials. And if we changed the gesture even slightly, they fell apart again.

At first, we thought chimpanzees had trouble using our gestures because there was something wrong with our tests. But because chimpanzees seemed to understand our intentions when they were competitive, but not when they were cooperative, we realized their failure might be meaningful.

In human babies, this is the spark that suddenly ignites, always early, always around the same age, and always before we can speak or use simple tools.[3] The simple gesture of extending an arm and index finger that we start to use at nine months old, or our ability to follow along when our mothers point to a lost toy, or a bird flying overhead, is something chimpanzees do not do and do not understand.[2]

This star of cooperative communication, missing from the constellation of abilities that comprise chimpanzee theory of mind, is the first to appear in humans.[5, 6] It shows up before we speak our first words or learn our names; before we understand that others can feel sad even while we are happy, and the other way around; before we can do something bad and lie about it, or understand that we might love someone and they might not love us back.

This ability allows us to communicate with the minds of others. It is the door into a new social and cultural world where we inherit the knowledge of generations. Everything we are as *Homo sapiens* begins with this star. And like many powerful phenomena, it begins in an ordinary way, with a baby understanding the intentions behind her parents' gestures.

If understanding these cooperative intentions is fundamental to the development of everything human, figuring out how that ability evolved could help us solve a major part of the puzzle of human evolution.

As Mike and I were discussing this one day, I blurted, "I think my dog can do that."

"Sure." Mike leaned back in his chair, amused. "Everybody's dog can do calculus."

* * *

It was reasonable for Mike to be skeptical. It was hard to be impressed with animals who drank out of the toilet and tangled their leashes around lampposts. Psychologists did not think dogs were interesting, so there was almost no research on their cognition. From 1950 to 1998, there were only two major experiments on dog intelligence, and both found that dogs were unremarkable. "Strangely enough," wrote one of the authors, "domestication does not seem to have produced anything new in dog behavior."[7] Everyone's attention was on primates. It made sense to study our primate relatives, who looked more like us and whose minds were presumably more like ours too.

Because people tended to assume that domestication made animals unintelligent, researchers looking for cognitive flexibility in nonhuman animals thought it best to look in the wild, where their survival depended on solving problems. How cognitively flexible could you be if you never had to think for yourself—if your food, shelter, and reproduction were all taken care of? But I knew my Oreo.

"No, really, I bet he could pass the gesture tests."

"Okay," said Mike, humoring me. "Why don't you pilot an experiment?"

GOOD DOG

Oreo's special talent was that he could hold three tennis balls in his mouth. When we played fetch, I often threw two or three balls in different directions. After Oreo fetched one, he would look at me to see where I had thrown the second ball.

I would point to it, and after he got that ball, he would look back at me again, and I would point to ball number three.

To show Mike what I was talking about, I took Oreo to play fetch.

"Hey, buddy, let's go."

He thumped his tail, a tennis ball in his mouth. When Oreo figured out where we were going, he started to sprint like a dog half his age. In our neighborhood, there was a large pond where Oreo and I used to play together.

Oreo bolted straight to the water's edge and barked the bark that said he would bark forever unless I threw the ball.

"Okay, okay. Just wait!"

I pulled a giant VHS video camera out of my bag and turned it on. I threw a ball into the middle of the pond and Oreo leaped after it. For a magical moment, he soared over the water; weightless, timeless, legs sprawled front and back, tongue lolling from his smiling mouth.

The splash was epic, as always. When Oreo got the ball, he came swimming toward me. I extended my arm and pointed to the left, but this time I had not thrown any other balls for him to find.

Oreo, failing to find a ball when he swam to my left, looked at me. I pointed to the right. He swam to the right. No ball. Then I called him in, took the ball out of his mouth, and threw it again, and repeated the pointing game ten times, so that Mike would see that Oreo's responses were not due to chance.

Mike watched the footage silently. Then he rewound it and watched it again.

I waited, nervously.

"Wow."

His eyes were bright with excitement.

"Let's *really* do some experiments."

The same behavior from two different individuals can be produced by two very different minds that understand the world differently. To attribute complex cognition you must follow the principle of parsimony:[8] You cannot infer complexity until all plausible simpler explanations have been ruled out. Experiments give us a way to do this.

Mike taught me that when you are probing the mind of someone who cannot talk, simple is best. Experiments are just a way of asking questions. If the question is easy to understand, it is likely the answer will be too. I called it "duct tape science"—if your equipment broke and could not be fixed with duct tape, your experiment was too complicated.

Even with just two cups, a table, and some duct tape, however, the experiments with the chimpanzees had taken months. Suiting up, waiting, food preparation, equipment checks, driving to see them, filling out forms, more waiting.

With Oreo, I took two cups and put them upside down on the ground a few feet apart.

"Sit."

I hid a piece of food under one of the cups. Then I pointed at the cup that hid the food. Oreo found it the first time. And the next seventeen times.

"Oreo," I said, scratching his ears as he hugged me with his full weight against my legs, "you're a genius."

All those months gesturing for chimpanzees and coming up with nothing, and Oreo had been sitting in my backyard the whole time, waiting for me to give him a chance.

Oreo and I were spending time together in a new way. Through my experimental games, I would give him a choice, and with each choice, he told me a little bit more about what the world was like for him. When I wanted to ask him whether he was really following my gesture or could just smell the food under the cup, I hid the food the same way but did not gesture. When he made his choice, he found the food only half the time. Without my help, he was just guessing. This meant that even though, like all dogs, his sense of smell was excellent, he could not rely on it to go to the correct cup on his first try.

It was lucky that Oreo and I were having fun, because working with him raised questions that involved a dozen variations on our game. Just because Oreo followed my point did not mean he understood the intention behind it, as a child does. There were simple explanations that could explain Oreo's success, and Mike helped me design experiments to test each one.

The most obvious was that Oreo was just following the movement of my arm, the same way he might watch a raindrop rolling down a window. He would not have to think the raindrop was trying to tell him anything to follow its path with his eyes.

The motion of my arm could have grabbed Oreo's attention as I pointed. As his gaze followed my arm, he could have searched for food in the cup he happened to be looking at—perhaps even forgetting there was another cup. This would mean Oreo did not understand anything about what I was thinking. I could have wiggled my arm or flashed a light on the same side as the correct cup and gotten the same results.

To control for this, I had to take the movement out of pointing. So sometimes I only turned my head and looked at the correct cup, other times I pointed across my body with the arm opposite the correct cup, and sometimes I got my little brother to cover Oreo's eyes until my arm was already extended and motionlessly pointing. In the hardest version, I even stepped toward the incorrect cup while pointing at the correct cup. Oreo had no problem finding the food in any of these new situations. He clearly was not just relying on my arm motion.

Oreo hadn't learned how to follow a point by trial and error like the chimpanzees. If he had been, he should have gotten better as we did more trials. Instead, he never made a mistake in the basic tests, and even in the more difficult tests, he did as well at the start as he did at the end. Whatever Oreo was doing, it appeared to be more flexible and cognitively sophisticated than the responses of chimpanzees.[9]

It was time to go bigger.

* * *

Oreo and I had grown up together. He might have learned to follow only my gestures. But could other dogs follow my point? I went to a doggy daycare in Atlanta, gathered dogs, hid the food under one of two cups, and gestured to the correct cup. Even though we had just met, the daycare dogs were just as good as Oreo at following my pointing gesture. Using this gesture seemed to be something all pet dogs could do.[10]

What makes human babies special is that they truly understand what you are trying to communicate with a pointing gesture, which means any gesture will do, as long as it's helpful. To demonstrate this with human mothers and babies, Mike asked the baby's mother to put a block on the correct cup. The babies had never seen their mothers do this before, but, guessing that she was trying to help them, chose the cup she put the block on. When I played this same game with dogs, they performed in the same way. Just like babies, they understood I wanted to help them and would use any new gesture they thought was intentionally helpful.[11]

Both dogs and babies were more likely to pay attention if you made eye contact and used a friendly voice. They could even use the direction of your voice. Human babies start to recognize vocal direction around their first birthday as they begin to understand that words refer to specific objects and actions. This might be why some dogs have been so successful at inferring the meaning of new words without any trial-and-error training.[12, 13]

Even the chimpanzees who were able to learn to follow

pointing gestures after dozens of trials could not generalize this skill when using a novel gesture—marking with a wooden block, for example, to indicate where the food was hidden. If we played a game of fetch with chimpanzees and extended an arm, pointing at a toy we wanted them to pick up, chimpanzees would bring back a toy, but not necessarily the one we pointed at.[14] They seemed to know only that pointing meant "Go pick something up and bring it to me." Instead of making eye contact with humans, as dogs do, chimpanzees spend more time looking at people's mouths.[15] This might explain their failure to be guided by our gestures.

We recently discovered that performance on related problems clusters in human infants.[16] Babies who understand what you mean when you reach toward the correct cup also understand when you gesture or look at the correct cup. Babies who struggle to understand pointing gestures also struggle to read the other types of gestures. But doing well on games that test communicative intentions does not mean they will do well in everything. A baby who is good at reading gestures is not necessarily good at physics, with a good sense of whether objects fall down or up, or which tool would work best to solve a certain problem. Those abilities are in a separate cluster.

We found that communicative intention skills cluster even more closely in dogs. If a dog does well on one gesture game, they do well on all of them. If they do badly on one, they do badly on all of them. Like babies, these skills do not relate to skills for solving problems that are not social. Not only do dogs have the suite of communicative intention skills that we have, but these skills cluster the same way. This means they share with us a specialized cognition for cooperative communication. Dogs look like us exactly where it counts.

Chimpanzees do not. Unlike dogs and babies, there is no relationship between chimpanzees' ability to use different communicative gestures. And unlike dogs and babies, their performance with different gestures is as likely to be related to nonsocial tasks as it is to be related to performance with other gestures. This means chimpanzees do not have the signal of specialized cognition. Instead, when they are solving these problems, they are using some general ability. Dogs and people are built for cooperative communication. Chimpanzees are not.[16]

Because cognition evolves to promote reproductive success, an animal will develop the most cognitive flexibility in the types of thinking that solve problems central to its survival. Unlike chimpanzees, dogs survive by communicating with people. But even I was surprised at how sophisticated dogs are in their understanding of our communicative intentions. How could dogs have social skills so similar to those that psychologists had thought were unique to us?

One of the obvious explanations was that something had happened during dog domestication that caused canine cognition to evolve. If this was true, and we could determine what it was, perhaps we could uncover what drove the evolution of cooperative communication not just in dogs, but in ourselves. Just like legs, eyes, and wings, which have evolved many times independently,[17] the ability to cooperatively communicate may also have evolved multiple times. Dogs may have converged with us cognitively in a narrow but crucial way.

Since dogs evolved from wolves, they have evolved to be more like us in a variety of ways. The gene that enables humans to digest starch also evolved in dogs, allowing them, unlike their wolf ancestors, to easily digest foods that humans gathered or farmed;[18] the gene that evolved to allow people to live at high altitudes also evolved in the Tibetan mastiff—both populations are able to harvest the low levels of oxygen found at high altitudes;[19] and the same gene that confers a degree of protection against malaria in West Africans has also been selected for in the region's domestic dogs.[20]

How did this convergence happen? Had we just chosen to domesticate wolves who already had some combination of these traits?

This was plausible but difficult to test. I didn't have the time to breed generations of wolves based on their ability to cooperatively communicate and see if they would turn into dogs. Without knowing more about how domestication happened, we could go no further with our investigation.

2

The Power of Friendliness

During Stalin's Great Terror of 1937–38, Nikolai Belyaev was arrested by the secret police for being a geneticist and shot without a trial.[1] Though Stalin was broadly paranoid about everyone, he disliked geneticists in particular because they seemed to go against the party line by promoting the popularized idea of the survival of the fittest. He saw this as an inherently American capitalist idea, a justification for those with superior strength or intelligence to amass wealth while workers lived in poverty. Stalin's solution was to ban genetics entirely. Genetics was taken off the curriculum of schools and universities. It was torn from textbooks. Geneticists were declared enemies of the state, sent to labor camps, or killed, as Belyaev had been.

A year later, Nikolai's brother, Dmitry Belyaev, became a geneticist too. In 1948, Dmitry was fired from the Central Research Laboratory of Fur Breeding in Moscow. But he kept his head down and in 1959 moved to Novosibirsk, as far from the political center of Moscow as you can get.[2] It was at this safe distance that he conducted the greatest behavioral genetic experiment of the twentieth century.

Belyaev's aim was ambitious. Instead of guessing how animals had been domesticated, he decided to domesticate them from scratch and see for himself. He chose to work with foxes, close but undomesticated relatives of dogs. The foxes' handlers had to wear mitts two inches thick because the foxes would struggle and bite when they were handled, but these animals were the perfect screen—breeding them for fur was important to the Russian economy and would ward off any suspicious government officials.

It was an elegant experiment. Lyudmila Trut, Belyaev's protégée, divided a population of foxes into two groups. She kept them in identical conditions but used a single criterion to separate them. The first group was bred based on how they reacted to people. When these foxes were seven months old, Lyudmila stood in front of them and gently tried to touch them. If the fox approached or was unafraid, it was chosen to breed with another fox who had a similar response. Only the friendliest foxes were chosen from each generation, and this group became the friendly foxes. The second group was bred randomly in regard to their response to people. Any differences between the two groups could be caused only by their selection criteria—friendliness toward people.[2, 3]

Belyaev continued this experiment for the rest of his life, and Lyudmila carried on after he died. By the time I arrived in Siberia, forty-four years after the work began, the regular foxes were pretty much the same as their ancestors. The friendly foxes were extraordinary.

* * *

Darwin was fascinated by domestication and used it to demonstrate the main principles of his evolutionary theory. After publishing *The Origin of Species,* he wrote *The Variations of Plants and Animals Under Domestication,* which used artificial selection to illustrate the way natural selection might work on

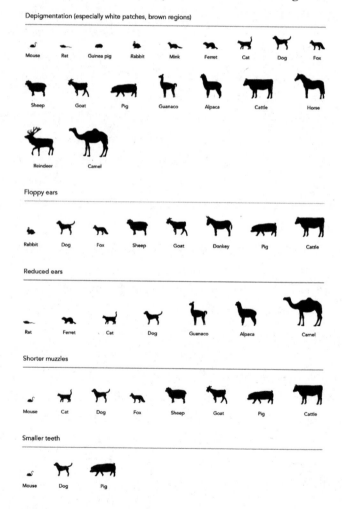

various inherited traits. But he did not offer a theory for when, where, and how animals first became domesticated.

Domestication has often been defined by physical appearances. Body size is a variable trait—in dogs, this leads to dwarf breeds like Chihuahuas and giant breeds like Great Danes. Dogs tend to have smaller heads, shorter snouts, and smaller canine teeth than their wild cousins. Their hair

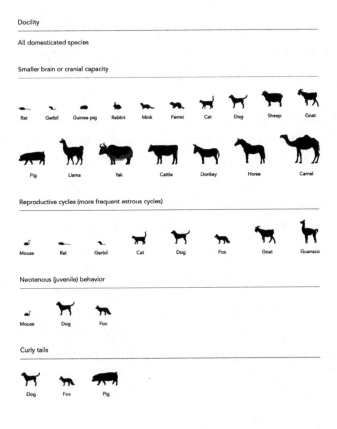

Docility

All domesticated species

Smaller brain or cranial capacity

Rat Gerbil Guinea pig Rabbit Mink Ferret Cat Dog Sheep Goat

Pig Llama Yak Cattle Donkey Horse Camel

Reproductive cycles (more frequent estrous cycles)

Mouse Rat Gerbil Cat Dog Fox Goat Guanaco

Neotenous (juvenile) behavior

Mouse Dog Fox

Curly tails

Dog Fox Pig

changes color so that they lose their natural camouflage. They can be covered in random splotches of color, sometimes with a star mutation on the forehead. Their tails curl upward, sometimes in a full circle like huskies' and sometimes in several loops, as in the tails of domesticated pigs. Dogs often have floppy ears. Instead of having one breeding season, they can breed throughout the year. This collection of traits is not unique to dogs; an assortment of these traits pops up in each domesticated species.[4]

Nobody knew what connected these seemingly random traits, or whether they were connected at all. Some thought people bred for these changes deliberately. The biologist Eitan Tchernov thought smaller animals would have been easier to handle and would have required less food.[5] The geneticist Leif Andersson said that farmers bred animals with splotchy coats to make them easier to spot when they wandered off.[6] The zoologist Helmut Hemmer said domesticated animals had weaker visual and sensory systems that lowered exploratory behavior, stress, and fear responses.[7] The benefits of having smaller teeth and being more fertile are obvious. But everyone tended to look at each trait associated with domestication individually, and many considered them deleterious. Most scientists had a low opinion of domesticated animals' intelligence, for example. As Jared Diamond wrote, brains presumably got smaller in domesticated animals because brains were a "waste of energy in the barnyard."[8] Everyone tends to agree that we deliberately chose to breed "animals more useful to humans than other individuals of the same species."[9]

Of the world's 147 large mammals (average weight over 100 pounds) with the potential to be domesticated, only fourteen were domesticated, and we relied on only five of those for any length of time: sheep, goats, cows, pigs, and horses. Smaller animals were also domesticated—wolves were one of these—but they were still relatively few.

Researchers proposed a set of conditions that would pre-dispose animals to domestication. Diamond proposed that they had to be able to eat food easily supplied by people, grow quickly, breed easily and give birth frequently in captiv-ity, have a friendly disposition, be predisposed to dominance hierarchies, and remain calm in enclosures or when faced with predators.[8] Diamond insisted that to qualify for domes-tication, an animal had to meet *all* these criteria. Other researchers added that suitable animals should be polyga-mous, readily controlled within a small home range, and that females should be able to live in large groups with males.

According to what became the dominant theory, domes-tication has always been a humancentric process by which animals come under our control and become economically useful. This theory explained, from a cultural and economic perspective if not a biological one, why particular animals had been domesticated and why some societies developed agriculture while others remained hunter-gatherers. But it had one big problem—dogs. Clearly dogs are domesticated, but their wild relatives, wolves, do not fit the essential crite-ria. Wolf food is difficult for people to supply. Wolves defi-nitely panic in enclosures, and while they are not normally aggressive toward humans, they do bite when threatened.

Belyaev thought domestication relied on just one criterion, and his theory promised to provide an answer to the question that had eluded everyone from Darwin to Diamond.

* * *

The friendly foxes were beautiful and strange. They had the grace of cats, but they barked like dogs. Some had black and white patches and blue eyes, like border collies. Some were spotted, like Dalmatians. Others were red, white, and black, like beagles. As Lyudmila led me around the site, all of them stood up and ran toward me, wagging their tails, whimpering and barking in excitement.

Lyudmila opened the door to one of their houses, and a russet vixen with black socks and a white star on her forehead leaped into my arms, licked my face, and peed for joy.

One of the first changes to the population of friendly foxes was coat color. Tawny red began to appear more frequently in their fur, then black and white splotching. After twenty generations, most of the friendly foxes were easily identifiable. A white star appeared on a few foreheads, then suddenly stars became frequent. Next came more floppy ears and curly tails. The friendly foxes had smaller teeth in shorter snouts, while the skulls of males and females became more similar to each other in their shape. These same changes had appeared in dog skulls early in domestication.[10, 11]

It was not just the foxes' appearance that had changed. Regular foxes bred only once a year. Many friendly foxes had a longer mating season. Some of the friendly foxes began to have two reproductive cycles, which meant they could mate eight months out of the year. They became sexually mature a month younger and had larger litters than the regular foxes.

Like wolves, the regular foxes had only a short window after birth—from sixteen days to six weeks—to become socialized to people. Like dogs, the friendly foxes have an expanded window of socialization that begins at fourteen days and closes at ten weeks.[10] Corticosteroids, or stress hormones, increased in the regular foxes between two and four months and reached adult levels by eight months. The friendlier the fox, the longer this surge in corticosteroids was delayed, and after twelve generations, the level of corticosteroids in the friendly foxes had been cut in half. In thirty generations, it halved again. After fifty generations, the friendly foxes had five times more serotonin—a neurotransmitter associated with lower predatory and defensive aggression—in their brains than the regular foxes.

In order to show that the changes were genetic, Belyaev and Lyudmila switched friendly fox kits at birth to see if they were influenced by their new mother's behavior. They implanted the embryos of friendly foxes into the wombs of regular foxes and vice versa. But it did not matter which mother gave birth to them or raised them. The friendly foxes were friendlier than regular foxes from the time of conception.[12]

The geneticist Anna Kukekova has already isolated a gene involved in the expression of friendly and aggressive behavior on the VVU12 chromosome, which is similar to the genomic region implicated in the domestication of dogs.[13] Other researchers have identified genes altered in the foxes and dogs that lead to Williams syndrome in humans, a condition characterized by hyperfriendliness.[14, 15] Future genomic comparisons will pinpoint exactly which genes were under selection to create the friendly foxes.

The genius of Belyaev's experiment wasn't to show that selection for friendliness created people-loving foxes but to reveal what came along for the ride. Floppy ears, shorter snouts, curled tails, splotchy coats, and smaller teeth were never intentionally selected for breeding, yet with each generation these traits became more common. Generation after generation, Lyudmila and her team selected only for friendliness and watched the physiological and physical changes accrue.[16]

Scientists have been able to replicate this experiment with species as distantly related from dogs as the chicken. Researchers bred a population of red jungle fowl—a wild Asian species from which all domesticated chickens evolved—for friendliness (that is, willingness to allow people to approach or touch them) and compared them to a control population. Just as Belyaev had predicted, selection for friendliness had, after only eight generations, made the experimental jungle fowl less fearful of novel objects, given them higher levels of serotonin and loss of pigmentation, made their bodies larger and their brains smaller, and made them more fertile.[17]

Dmitry and Lyudmila had done in one lifetime what it typically takes nature thousands of generations to accomplish, and they had a formula to show for it: Domestication occurs when animals who are friendly toward people become more successful at reproducing.

My graduate adviser at Harvard was Richard Wrangham. When I discussed the Russian foxes that looked like dogs with him, he saw even deeper implications. A population of fearful and aggressive foxes, selected only to be attracted to humans, in a few generations started displaying accidental changes that were not selected for: Could a change in cognition be another accidental change?

What Richard was suggesting seemed impossible. We were not talking about floppy ears or a curly tail. Reading cooperative communicative intentions was one of the most critical aspects of the theory of mind to appear in human babies. It made sense that dogs who were better at reading cooperative communicative intentions would have been more successful reproducing as a direct result of this enhanced skill. Or was it, as Richard suspected, a trait that accidentally evolved like splotchy coats did? No one had ever tested something like this before. So Richard convinced me to travel to Siberia to test the foxes myself.

There were just a few minor problems with the Siberian plan. I had never seen a fox in my life. I did not speak a word of Russian. No one had ever tested the cognition of any fox, anywhere. It had taken me years to test theory of mind in chimpanzees and over a year to test the dogs. To test these foxes, I had eleven weeks. The regular adult foxes were scared of people. And I had to test both groups. The only thing on my side was timing. I happened to arrive at the end of spring when the farm was full of baby foxes.

Richard sent me with one of his undergraduate students, Natalie Ignacio, and I put her on full-time hugging duty. She had to socialize a group of regular fox kits. At just a few weeks old, they had not developed their adult fear system. Natalie squealed as she crouched among the dozen silver furballs sniffing her curiously.

"Do whatever it takes to make them adore you," I pleaded. "They have to be ready to test in two months."

I left Natalie and trudged across the farm to the three- and four-month-old foxes who had just been separated from their litters and put into two groups of six regular foxes and six friendly foxes. Each group had its own enclosure. I sat between them, looking inside their pens.

As soon as I sat down, the friendly foxes started to squeak and whine. They panted and scratched their door, wagging their tails. When I scratched their ears, they licked my hand, then lay back and closed their eyes so I could rub their bellies. When I gestured, they closely followed the motion of my hands.

The regular foxes watched me steadily. I did not make any sudden movements or loud noises. I did not try to touch them or play. I just watched and waited. They hid in the back corner of their rooms. For my experiment to work, I needed to find something that could tempt both groups of foxes long enough to get their attention.

Weeks went by with no success. Then an answer came from the sky. A hawk with a four-foot wingspan swooped over the fox houses. The foxes watched the hawk, enraptured. Then a single wing feather fell spinning to the ground, and all the foxes, from both groups, charged over to stare at it.

The next day, I picked up a feather on my way over to the foxes.

"You guys like feathers?"

Every fox eye was on me. I wiggled the feather in front of a regular fox. Instead of scampering to the back of her pen as she usually did, she padded toward me and batted the feather. The friendly foxes did the same thing.

Bingo. I had something all the foxes were equally attracted to. I wiggled a feather in front of a fox until she was standing in front of me. Then I pointed to one of two toys. I pushed the two toys toward the fox and recorded which toy the fox played with.

The regular foxes would play with one of the toys, but they did not choose the toy I gestured at. They just chose randomly. The friendly foxes preferred to play with the toy I'd suggested. Even though the regular foxes and friendly foxes had spent the same amount of time with me, only the friendly foxes followed my gestures.

After nine weeks of hugging and training, Natalie had a group of regular fox kits who could find food under the bowls. It was time for the test.

Natalie hid the food under one of two bowls and pointed to the food. Like chimpanzees and wolves, the regular foxes' results were barely above chance. They were mostly guessing.

Then we tested the friendly fox kits whom Natalie had never met. She just turned up at their enclosures, got them out, and hid food under one of two bowls. If humans had selected dogs only for their cooperative communication skills, the friendly foxes, selected only for the trait of friendliness, would not have the cooperative communicative skills needed to follow my gestures. But they did. The friendly foxes were not only as good as puppies. They were a little bit *better*.

Richard was right. Without ever having played this kind

of game before, the friendly foxes could use our gestures to find the food, just like dogs, while even after months of intensive socialization, the regular foxes were barely above chance when we gestured.[18]

If you want a smarter fox, breed the friendliest foxes you can find. Wild foxes already had the ability to respond to the social behavior of other foxes. Belyaev bred the foxes to decrease their fear of people, likely allowing an evolutionarily ancient social skill they use in interactions with one another to flourish in a new context, in a relationship with humans.

Unconstrained by fear, foxes could use social skills, such as cooperative communication, more flexibly. Problems that were previously confronted alone became social problems that were easily solved with cooperative partners. Cooperative communication had been enhanced, but contrary to what most hypotheses of cognitive evolution predicted, it had been an accident. This kind of social intelligence is just another side effect of fearfulness being replaced by friendliness.[19] The fox work provided strong evidence that the basic skill behind the cooperative communication we observed in dogs was the product of domestication.

We had also discovered that this skill we found in dogs was not simply the product of an individual dog interacting with humans for hundreds, if not thousands, of hours before they were adults. When we tested puppies of different ages and rearing histories, we found that even the youngest puppies were excellent at understanding human gestures. In fact, puppies between six and nine weeks old scored perfectly in experiments using basic pointing gestures as well as novel gestures they had never seen before. [20, 21, 22] This is impressive because at six weeks old, puppies have underdeveloped brains and are still learning to walk.[23] And their aptitude went beyond visual gestures. Puppies could also use the direction of a person's voice to find food—they were even better at using this vocal pointing than adult dogs.[24] All the cooperative communicative skills of dogs are already present in puppies. They are only enhanced through interactions with humans. In animals, it is rare to see such a flexible cognitive ability come online so early without extensive experience. Reading human gestures seems to be one of the first social skills to appear not only in human babies but in dogs.

We also learned that dogs did not simply inherit from their wolf ancestors their ability to cooperate and communicate with humans. Wolves must be skillful at reading the signals of one another and their prey, so it seemed plausible that these skills might easily generalize to their interactions with people.[25-27] Just as we did with dogs, we hid food treats under one of two cups and gestured to the correct cup to help the wolves find the food. But when it came to following human gestures, the wolves we tested looked like chimpanzees.[22] Even after dozens of repetitions, the wolves were still guessing. We had found that puppies with barely any human contact did better at reading our gestures than adult wolves. As smart as wolves are, they were unable to spontaneously understand human cooperative communicative intentions.[28]

Researchers continue to compare wolves and dogs to better understand how experience and training may also shape differences between the two species,[29-33] but as with the experimental foxes, the unusual cooperative communicative abilities of dogs evolved as a result of domestication.

A WOLF AT THE DOOR

According to Google, "How to stop my puppy from eating poop" was one of the top ten searches about dogs in 2015.[34]

But poop is central to the story of how dogs came into our lives.[35]

When we arrived with our projectile weapons to hunt and gather in Eurasia ~50,000 years ago, we wiped out nearly every predator of the Ice Age except the wolf.[36]

People generally assume that thousands of years ago, agriculturalists picked up some wolf puppies, brought them home, and bred the tame wolf puppies to make tamer wolves. After many generations, we had created the lovable dog. Based on genetic work, we know this cannot have happened, since wolf domestication began at least ten thousand years before the first seed was sown. Hunter-gatherers would have been the first people to live alongside the first dogs.[37]

Intentional domestication of Ice Age wolves evokes an impractical scenario. Humans would have had to allow only the friendliest, least aggressive wolves to reproduce over dozens of generations. This would mean that for hundreds of years, if not longer, hunter-gatherers would have lived among these large, impulsively aggressive wolves, sharing many pounds of hard-won meat with adult wolves every day. More likely, there was a stage in domestication before humans took control—a period of self-domestication.[38]

If we created anything, it was lots of garbage. Even today, hunter-gatherers dispose of food waste and urinate and defecate outside camp. As our population became more sedentary, there would have been more tempting morsels for a hungry wolf to eat at night. Discarded bones would be nice, but because we cook and rapidly digest our food, human feces is equally nutritious.[39] Our waste would have been irresistible to any wolf who was calm and brave enough to approach our camps. These wolves would have been at a reproductive advantage and, scavenging together, more likely to breed together. The flow of genes between the friendly wolves and the fearful wolves would have subsided, and a new, friendlier species could have evolved even without intentional selection by humans.

After only a few generations of selection for friendliness, this special population of wolves would have begun to look different. Coat color, ears, tails: All probably started to change. We would have become increasingly tolerant of these odd-looking scavenger wolves and would quickly have discovered that these protodogs had a unique capacity for reading our gestures.

Wolves could understand and respond to the social gestures of other wolves, but they were too busy running away from humans to pay attention to our gestures. Once their fear of us was replaced with attraction, their social skills could be used to communicate with us in a new way. Animals who could respond to our gestures and voices would be extremely useful as hunting partners and guards. They would have been valuable also for their warmth and companionship,[40, 41] and, slowly, we would have allowed them to move from outside our camps to our firesides. We did not domesticate dogs. The friendliest wolves domesticated themselves.[2]

These friendly wolves became one of the most successful species on the planet. Their descendants now number in the tens of millions and live with us on every continent, while the few remaining wild wolf populations, sadly, live under constant threat of extinction.

If self-domestication could happen in dogs without interference from us, then what about other animals, especially those who, like those early wolves, are encroaching upon human habitat today?

Like protodogs thousands of years ago, urban coyotes scavenge our trash, with human garbage making up to 30 percent of their diet.[42] These urban coyotes raise their pups in drainage ditches, under fences, and in pipes. They cross freeways with traffic volumes of more than a hundred thousand cars per day, and casually stroll over bridges like pedestrians.[42]

My student James Brooks and I analyzed the data from camera traps all over North Carolina.[43] We predicted that when we coded coyote behavior toward the camera, we would see a link between temperament and human population density. Our initial results suggest that urban coyotes are more likely to approach trap cameras than coyotes living in wild areas. It is not just their temperament that makes coyotes adaptable. When we compared the self-control of thirty-six different species, coyotes were not only better than dogs and wolves, they were the only other animal who were as good as great apes.[44]

Red foxes in the United Kingdom are at densities ten times higher in urban areas than rural. Urban arctic foxes are more likely to breed earlier, as yearlings.[45] Urban blackbirds in Europe are less aggressive than their rural relatives. They have higher breeding density and a longer breeding season.[46] They also live longer and have lower corticosterone, a stress hormone, than their rural relatives.[47]

In the Florida Keys, there is a population of deer that is native to the islands called Key deer. These deer, who live in more urban areas, have become less fearful, bigger, more social, and more fertile than those out of contact with humans.[48] In other urban areas, people have seen common white-tailed deer with unusual coloring—like splotchy or albino coats. There are anecdotes of "deformities" in piebald and albino deer, including short legs, shorter, lower mandibles, and longer tails—the kind of changes associated with the domestication syndrome.[49]

After we uncovered how cognitively sophisticated dogs are, other researchers began to reevaluate assumptions about the intelligence of domesticated animals more generally. Researchers have found increasing evidence that friendliness, rather than dulling intelligence, gives animals a cognitive advantage—particularly when it comes to cooperating and communicating.

József Topál found that domesticated ferrets are better at following our gestures than wild ferrets. This is striking since, unlike many dog breeds, domesticated ferrets do not cooperatively communicate with humans during their traditional hunting roles (i.e., ferreting out rodents). It again suggests that increased skill at reading human gestures came along with more friendliness toward people since humans had no motivation to intentionally breed ferrets based on this trait. [50]

Kazuo Okanoya compared white-rumped munia to domesticated Bengalese finches. He found that the Bengalese finches were less aggressive than the munia finches. They were also less stressed, with lower corticosteroids in their feces than munia finches, and less fearful of new objects. Surprisingly, Okanoya found that the songs of the domesticated Bengalese finches were more complex than those of the munias. The Bengalese finches could also learn multiple songs from other birds, but the munia finches could learn their simpler songs only from their fathers. When the birds were cross-fostered, the Bengalese finches could easily mimic the munia songs, but the munias could never master the more sophisticated Bengalese songs.[51]

In 2008, the human population turned a corner. There are more of us living in urban areas than rural ones. We have become an urban species.[52] By 2030, the three billion humans in cities will become five billion.[52]

Unlike other models that suggest domestication can occur only in rare species that are useful to humans, Belyaev's work predicts that our increasing population density will be enough to drive the next great self-domestication event through natural selection. It could happen very quickly, depending on the strength of the selection pressure, the size of the initial population, and how genetically isolated the population is from wild populations. Any species capable of exploiting human spaces by replacing fear with attraction will not only survive but thrive.

3

Our Long-Lost Cousins

If dogs and other urban animals domesticated themselves by becoming friendlier and more attracted to humans, we began to wonder if the same thing could happen if we took humans out of the equation. Could an animal become self-domesticated when interacting with others of its own species through natural selection?

Few animals are friendlier than the bonobo, but bonobos have always been a puzzle. Bonobos and chimpanzees shared a common ancestor around a million years ago and share more genes with us than they do with gorillas. This makes bonobos and chimpanzees our two closest living primate relatives. It is like having two first cousins who are equally related to you. They are similar but different from each other in important ways.

Some of the ways bonobos differ from chimpanzees have been difficult to explain, until we realized they mirrored changes we see in domesticated animals. Male bonobos' brains are approximately 20 percent smaller than those of male chimpanzees, and bonobos of both sexes have smaller faces with smaller, more crowded teeth. Some bonobos' lips are missing pigment, giving them a pinkish tinge. They are also missing pigment in their tuft, a patch of long spiky hair on their bottom. Chimpanzees have these tufts when they're young but lose them when they mature. Both bonobos and chimpanzees are playful when they're young, but chimpanzees grow out of this behavior, while bonobos carry this joie de vivre into adulthood. Bonobo adults are also more playfully sexual, and females use sex to bond with one another and to resolve conflicts among the males.[1]

People have tried to explain the function of these traits individually, just as they had with other domesticated animals. Richard Wrangham was the exception. He had not sent me to Siberia to find out how smart dogs were. He wanted me to find out what domestication had done to Belyaev's foxes because he thought it might explain what had happened to bonobos.[2]

I will always remember one terrible day at Yerkes, the primate research center where I studied chimpanzees in my early days with Mike Tomasello. It was a Sunday, and I was the only person there.

I was playing the gesture game with a chimpanzee called Tai. She was old and moved slowly, but, like an indulgent grandmother, she was humoring me. She watched me hide the food under one cup and point to it. Tai wrinkled her face, as though puzzling over a difficult crossword. She pointed. Got it wrong. Slapped her forehead.

Suddenly there was a scream so loud it shook the walls. Tai and I froze. I jumped up, knocking over the table and everything on it, and ran down the hall.

Travis, a male chimpanzee, was being held down by four other chimpanzees. One chimpanzee held his legs and two held his arms, so he was spread-eagled on his stomach.

An enormous toothless female called Sonia sat on his back, pinning him to the floor. Normally Sonia's weight was comical, but today it was frightening. Travis struggled beneath her, but he could not get up.

"Get off him!" I yelled as loud as I could, but my voice was lost in their shattering screeches.

The two chimpanzees holding Travis's arms took turns kicking his head into the concrete, screaming horribly. They had already bitten off two of his fingertips. His mother was nearby, wailing, but she was as helpless as I was.

I had seen chimpanzees fight before. I had seen them bite and hit each other. I had seen a chimpanzee casually break a woman's hand. But this was different. They were going to kill him.

In the wild, male chimpanzees regularly patrol the border of their territory. Before each border patrol, they huddle and put their arms around one another. They put their fingers in the others' mouths and touch the others' testicles as a sign of trust. They fall silent and walk single file. Every now and then they listen and sniff the ground near the border to see if the enemy has been nearby and how many of them there are.[3] If they outnumber the enemy three to one, they are more likely to attack. They pin their victims to the ground, bite bits of their fingers off, and even disarticulate limbs. In extreme cases researchers have found corpses with torn jugulars and missing testicles.[4, 5] If a chimpanzee community kills enough males from the neighboring territory, they move in, effectively annexing this new territory along with any females living within it.[6] Richard has noted that many human communities, from hunter-gatherers to street gangs in Chicago, conduct border patrols and raids in a similar manner, and he found that the homicide rate of chimpanzees and hunter-gatherers was similar.[7]

It is not just male chimpanzees who have the potential for violence. Female chimpanzees also have strict hierarchies illustrated by where they sit in a fruit tree. The high-ranking females sit at the sunny crown and eat the best fruit. Mid-ranking females have to make do with lower branches. The lowest-ranking females are pushed to the outskirts of the group's territory, leaving them vulnerable to attacks by neighboring males. When females reach puberty, they leave their mother's community and try to find mates elsewhere. Immigrating females are often beaten so badly by female members of a new group that the high-ranking males have to step in to prevent them from being seriously injured.[8]

I had not yet seen chimpanzees in the wild, but standing outside the enclosure at Yerkes, I knew Travis was in serious trouble. The floor was covered in a dark pool of his blood. More blood spurted from a gash in his thigh.

I grabbed the hose and turned it full blast on Sonia. She screamed at me in fury and jumped off Travis, who ran with his mother into another room. The rest of the chimpanzees were right behind him. I held them off with the hose and threw all my weight against the door, slamming it shut.

Travis collapsed, heaving, into his mother's arms while she carefully examined his wounds. Thankfully, Travis survived. But because the enclosures were so poorly designed, all the Yerkes managers could do to prevent more violence was to split the group in two. Once the groups were separated, even friends and family members would never touch one another again.

It was a clear demonstration of just how costly aggression can be. It can lead to serious injury or death, but it can also severely limit your number of social partners. The risks associated with aggression pay only when they result in more or higher-quality offspring. Any slight tweak to this cost-benefit ratio, and friendliness quickly becomes more advantageous than aggression.

Lola ya Bonobo is a hidden forest just outside the Democratic Republic of Congo's capital, Kinshasa. It is a haven of nature in a sprawling city of over 10 million. From the moment you step inside, you feel you are deep in the Congo Basin. There is a lake filled with lilies. Plants burst with flowers in the shape of birds.

As I walked along the forest path, a black bundle fell from the sky and wrapped her arms around my neck.

"Hey, Malou." She squeezed my waist with her feet. "Does your mama Yvonne know you're here?"

Malou laughed, and on cue, an annoyed voice cut through the morning.

"Malou! *Où es-tu?*"

Malou leaped from my back and disappeared into the trees.

Malou was found in the hand luggage of a Russian couple in the Paris airport, a victim of illegal pet trafficking. It was just before Christmas, and the X-ray operator saw something the size and shape of a small child curled in a fetal position at the bottom of the bag, covered in mangoes. Airport officials scratched their heads over what to do with the small creature. Her stomach was swollen and bloody. Her feet were covered in burns. A rope had been tied around her so tightly it had lacerated her groin. She was so dehydrated she could barely move.

She did not look as though she would make it through the night and would have been euthanized except that Claudine André, who founded Lola ya Bonobo, heard about Malou and came to her rescue. Through her contacts with the Environmental Ministry and the French embassy, she was able to reach then president Jacques Chirac, who had Malou flown home to Congo.

When orphans arrive at Lola ya Bonobo, a veterinarian treats their wounds. Then they are given to a woman who mothers them, or, if they are old enough, sent to a nursery with other orphans. When they are grown, they spend their days in a large forest with other bonobos, and then come into night buildings to sleep. Like us, they suffer but are also resilient. When Malou arrived, she was freezing and infested with parasites. Her hair fell out in clumps. Gently, Mama

Yvonne, her human caretaker with children of her own, nursed Malou back to health.

In any bonobo group, either in the wild or in captivity, there is never an alpha male. As a result, many scientists thought females were in charge.[9] No one suspected the important role babies play.

Infant chimpanzees are cautious taking food from anyone, especially big males. So considering babies in a chimpanzee group when assessing everyone's rank is not very informative. As we observed the bonobos at Lola, however, their natural behavior and interactions suggested something different was going on. To our surprise, we repeatedly saw adult males running away from food when infants were sitting nearby. My student Kara Walker and I decided to do systematic observations, but unlike in other studies, we also included the baby bonobos in the dominance equation of each group. When we did, some of the highest-ranking bonobos were babies who also had their mothers in the group. The babies who are raised by their mothers at Lola outrank some of the adult males. And even if an adult male is higher-ranking than a baby, the male is always very well behaved around it.[10] Watching a male bonobo run away from a baby bonobo the size of his foot seems ridiculous, but it makes sense if you look at it from the point of view of the baby's mother.

For a female, one of the worst things that can happen to her reproductive fitness is that someone kills her baby. Not only are her genes not passed on; babies are energetically expensive investments. When a female is pregnant and nursing, her body diverts an enormous number of calories to the baby. To have a belligerent male kill your baby with one savage swipe is a devastating loss on reproductive investment.

It would be a huge advantage for females to eliminate this risk. Female chimpanzees mate with several different males to confuse paternity and reduce the risk of infanticide. But female chimpanzees are betrayed by their own bodies. Males know when a female is ovulating because the pink part of her bottom swells, advertising exactly when she is most likely to conceive. All the high-ranking males attack an ovulating female, beating her into submission so she will mate with them and not other males. Her only defense is to stay close to the alpha male as protection, which means the alpha male is the most reproductively successful. It also means that if the alpha male loses his status while her infant is still young, the new alpha may attack her infant. This perpetuates a cycle of violence in which aggressive males have the advantage and infanticide can increase fitness in males by quickly returning nursing mothers to a reproductive state.[3, 11]

Female bonobos have managed to break this cycle by obscuring when they ovulate. They have swellings throughout their ovulatory cycle, making it harder for males to determine exactly when they are fertile. Females are also aggressive toward males who start acting like chimpanzees. Any male who tries to force females into mating is met with fierce opposition—sometimes from a coalition of angry females. And if any male even looks at a baby the wrong way, he quickly feels the full force of female wrath. Females work together, so that even though males might dominate by size, females always dominate by numbers.[11–13]

And while female chimpanzees help only their relatives, female bonobos help all females. When a new female arrives, she is treated with kindness and excitement. The other females rush to greet her, compete to groom and rub genitals with her. These resident females will defend the new female against males they have known for years. They will even defend her against their own sons.[14, 15]

Richard proposed that bonobo society evolved to be friendlier because food resources south of the Congo River were more predictable. Ecological studies suggest that fruit and herbs are more plentiful in bonobo forests. Nor do bonobos have to compete against gorillas for these resources. Gorillas often live in chimpanzee forests, but they do not live south of the Congo River, near bonobos.[3, 16]

Both high- and low-ranking bonobo females are able to meet their daily energy needs, while only higher-ranking chimpanzees are guaranteed enough food each day. Female bonobos can afford female friends, while chimpanzee females must compete. Friendly female bonobos can support one another and don't have to put up with male aggression. They also prefer to mate with the least aggressive males. For male bonobos, friendliness became the winning strategy.[1, 16]

No male bonobo has ever been seen to kill a baby. Male bonobos do not form gangs to patrol their border or commit lethal aggression against their neighbors. No bonobo in captivity or the wild has ever been observed to kill another bonobo.[17] In fact, neighboring groups of bonobos are just as likely to travel together, share food, and interact amicably as to show any hostility.[11, 18]

The female victory is so complete that the best chance for a male to get access to females is through his mother. Rather than forming coalitions to coerce females, as male chimpanzees do, male bonobos rely on their mothers to introduce them to female friends.[19] Chimpanzee males dominate their mothers, but bonobo males are the ultimate momma's boys.[20] This type of friendliness toward females is such a successful reproductive strategy that the most successful male bonobos are far more successful at fathering offspring than even the most successful alpha male chimpanzee.[21, 22] This finding supports the hypothesis that the female preference for male friendliness is the selection pressure that causes a friendlier society to evolve.[11]

Bonobo male

Chimpanzee male

Remember, wolves who were attracted to people had such an advantage that friendliness became a powerful selection pressure. This pressure caused behavioral, morphological, and even cognitive evolution. If a species, like the bonobo, undergoes natural selection for tolerance and friendliness, not toward people but toward their own kind, can this also lead to self-domestication?

Malou was fearless of authority. Even without her mother, she seemed to know that as a baby bonobo, she could pretty much do whatever she wanted. All the babies in the nursery were the same. We would hear a rustle in the trees during lunch, and a black bundle would drop onto the table, kicking over the dishes and grabbing handfuls of food on the way out. We would go into the kitchen to make a cup of tea and find baby bonobos rummaging through the drawers. One baby got into a tin of powdered milk and came out looking like a miniature abominable snowman. Another baby drank a whole bottle of dish soap and spent the afternoon burping bubbles. Baby bonobos are pure abandon and reckless joy.

If bonobos are self-domesticated, then we should be able to find in them the same markers of the self-domestication syndrome that we see in other domesticated animals. Based on the self-domestication hypothesis, there were predictions we could test by comparing bonobos and chimpanzees. Bonobos had some of the physical features of the self-domestication syndrome. But if bonobos really were self-domesticated, we should be able to experimentally demonstrate the following:

1. Bonobos should be more tolerant toward one another than chimpanzees, even in stressful situations.
2. Bonobos should have physiological mechanisms to prevent aggression.
3. As a by-product of their tolerance and friendlier physiology, bonobos should have more flexible cooperative communication skills than chimpanzees.

Our predictions neatly mirrored what we had tested in dogs and foxes. The problem was that no one had ever experimentally compared chimpanzees to bonobos before—on any test at all. Some scientists doubted that bonobos and chimpanzees were even different.[23] Lola ya Bonobo sanctuary, with its large bonobo population, provided the perfect opportunity to test our predictions.

Our first step was to see whether bonobos were more tolerant toward one another than chimpanzees. There is an easy way to test someone's tolerance. Just ask them to sit down and share a meal with someone else. So before the bonobos ate breakfast, we placed a single pile of fruit in a room, then released a bonobo into the room. This hungry bonobo could either eat all the food or could share the food with another bonobo by opening a one-way door separating their rooms. A chimpanzee would simply eat the food without opening the door. Remarkably, the bonobos opened the door and shared their food. Bonobos prefer to eat together even if it meant losing some food.[24]

We then created an even more complicated situation. We again released a hungry bonobo into a room with a nice pile of fruit, but this time they could choose *with whom* they shared food. They could share food with either one of their groupmates or a bonobo from a different group they had never met. Overwhelmingly, bonobos opened the one-way door between themselves and the stranger. They preferred sharing food and interacting with a new bonobo over someone they already knew well.[25] In additional tests, bonobos were also willing to help a strange bonobo even if they received no reward for helping.[26] With little reason to fear strangers, bonobos appear eager to start new friendships. Even the Good Samaritan might be impressed with their willingness to help a stranger in need.

This type of friendly interaction among strangers is unknown in chimpanzees. Adult male chimpanzees are more likely to be killed by a male stranger than by anything else.[5] But we found that bonobos are not only nonaggressive toward strangers, they are *attracted* to them. Bonobos are far more tolerant than chimpanzees.[11]

We also see further evidence of this difference in their physiological response to stress when sharing. Before we gave male bonobos the food sharing test, we found that their cortisol, the hormone related to stress, increased, likely in anticipation of a potential conflict over the food. When we looked at the hormones of male chimpanzees in response to sharing, we found a different response. In anticipation of potentially competing over the food, their testosterone rose rather than their cortisol. They were hormonally primed for competition.[27]

It seems likely that chimpanzee testosterone levels are higher even in the womb. In mammals, if a mother produces high levels of androgens (including testosterone) while she is pregnant, her baby's second digit (index finger, or 2D) will probably be shorter than the fourth digit (ring finger, or 4D). This ratio is called 2D:4D. When we measured the 2D:4D of chimpanzees and bonobos, we found that chimpanzees' second digit, or index finger, is in fact shorter than the ring finger. Bonobo fingers don't show this effect. This suggests that even before bonobos are born, they are already exposed to fewer of the hormones that masculinize chimpanzees.[28]

When the neuroscientist Chet Sherwood looked at bonobo amygdala, the part of the brain that reacts to threat, he found that bonobos have double the serotonergic axon density in the basal and central nucleus of the amygdala than chimpanzees.[29, 30] This means that the serotonin—the same neurohormone that changes in foxes and other animals as they are selected for friendliness—is altered in bonobos. In experimentally domesticated animals, a change in serotonin is among the first changes that accompany increased friendliness.[31, 32] What this means is that bonobos have physiological mechanisms to prevent aggression and promote friendliness, much as domesticated animals do.

Domestication can also affect communicative abilities. To see if bonobos' cooperative communication skills were more flexible than those of chimpanzees, we developed a cognitive battery of twenty-five games that we played with more than three hundred chimpanzees, bonobos, orangutans, and children. We found that chimpanzees and bonobos were similar in almost every cognitive test—except that bonobos were better than chimpanzees in games that assessed abilities relating to theory of mind. Bonobos were especially sensitive to the direction of a human's gaze.[33, 34]

Like domesticated Bengalese finches with their more complex call structures, bonobos also showed more vocal flexibility than chimpanzees. Bonobos frequently use "peep" vocalizations that can mean different things. Other bonobos have to use the context of the situation to infer the meaning of the peep, in a way that's similar to the way we learn language. This is not true for chimpanzees.[11, 35]

In order to test overall cooperation in bonobos and chimpanzees, we did a different test. We threaded a rope through loops on either side of a plank. Food was placed on the plank, and the plank was set out of reach. The only way to pull the plank forward was to pull on the rope at the same time as someone else (the ends of the rope were placed within reach of the apes but were too far apart for one ape to pull). If someone pulled too hard, or tried to pull it unaided, the rope came unthreaded and no one got the food. Success required cooperation.

We tested chimpanzees, and a few pairs of them were extraordinary. They spontaneously solved the problem on the first attempt. They knew when they needed help, they knew who was the better cooperator, and they could successfully negotiate—even without norms and language.[36, 37] But we could not take the chimpanzees from successful pairs and re-pair them with others. They were too intolerant of one another.[38]

Nor could chimpanzees share food unless it was divided into two piles. All it took for chimpanzee cooperation to fall apart was to set the food in one pile in the middle of the plank. One chimpanzee would end up eating all the food and the other chimpanzee would either quit or sabotage the game by pulling the rope out of the loops. Even though the same two chimpanzees had cooperated successfully before, they couldn't negotiate the division of that one pile of food.

Unlike the chimpanzees, who had months of practice and preparation for this test, bonobos could cooperate immediately. When we moved the food from two piles into one pile, they cooperated. When we mixed up the pairs, they cooperated. In all situations they ate together happily.[39] And not only did they share the food, but when someone reached the food first they left enough food for their partner, so that each ended up with half.

Bonobos had beaten the chimpanzees at solving this cooperative problem, even though they had been totally naïve

compared to the better-educated chimpanzees. Tolerance wins over knowledge when cooperation is required.[40]

A LIFE WITHOUT WAR

Self-domestication causes all sorts of changes. Some are endearing. Some are fascinating. Some are just bizarre. But the one change that links all the others, the change that happened first, that is present in every domesticated animal, is also the most important—an increase in friendliness.

Bonobos have been celebrated and mocked as the make-love-not-war hippie ape. They have been ignored as many have looked at the more familiar chimpanzees as a more suitable mirror for ourselves. After all, chimpanzees have almost everything we have. A light side and a dark. A bright intellect and devilish mischief. Tenderness and murder in the same breath.

But we ignore the bonobos' example at our peril. Among our great ape relatives, bonobos have escaped the lethal violence that plagues the rest of us. They do not kill one another. And that is a feat that, despite our intelligence, we have yet to accomplish.[11]

4

Domesticated Minds

Could we be self-domesticated? Could domestication account for our unique cognitive capabilities? At first glance, these ideas seemed far-fetched. As remarkable as dogs and bonobos are, the changes we observe in their evolution, from their common ancestor with wolves and chimpanzees respectively, seem dwarfed by the changes that must have occurred during the evolution of our species.

The more we learned about the way self-domestication impacts animal cognition, however, the more plausible these ideas became. After all, the evolution of cooperative communicative skills we observed in both dogs and bonobos represents the type of cognitive evolution in humans we need to explain. Fortunately, our knowledge of human development and neuroscience has advanced sufficiently for us to test the idea.

It was selection for the way Belyaev's foxes responded emotionally to humans—friendly or afraid—that shaped the way they were able to communicate. Does this link exist in humans? The psychologist Jerome Kagan, a pioneer in the study of human emotional reactivity, systematically measured how hundreds of people responded to new situations, objects, and people from the time they were babies until after they were in college. When he first tested the emotional reactivity of four-month-old babies, he found tremendous variation in how they responded. When the babies were introduced to something new, some were highly reactive, arching their backs and crying. Others were mildly reactive, calmly babbling and reaching to touch the strange objects. Kagan followed these babies for decades, testing them every few years. He found that the quality and intensity of their emotional reactions at four months old often predicted what they would be like as adults.[1]

Deep within each of our brain's two hemispheres sits an amygdala, a part of the brain activated by threats. Kagan predicted that, just as in animals, people's emotional reactivity would be influenced by the amygdala. Kagan found that not only was this the case, but also people's emotional reactivity corresponded to whether they had been high- or low-reactive as babies.[2]

Henry Wellman, a psychologist who read our work on domestication, became curious to see whether variation in emotional activity, like that Kagan had discovered, was related to the development of theory of mind in children. Like us, Wellman reasoned that if changes in emotional reactivity in dogs and foxes changed how they read the communicative intentions of others, perhaps the same relationship existed in human children.

One very sophisticated ability that emerges from theory of mind is false belief, which allows you to understand that what someone else thinks is mistaken. It usually does not fully come online until a child is four years old. Wellman found that shyer children with lower emotional reactivity developed an understanding of false belief more quickly than children with higher emotional reactivity.[3] Early understanding of false belief is also linked to early language development, so children with low reactivity had an advantage in both cooperation and communication. Low reactivity even seems to influence the speed at which we develop the ability to cooperate and communicate.[4-8]

Further support for this connection comes from the tentative brain map of the regions activated when we use theory of mind: the medial prefrontal cortex (mPFC), temporal parietal junction (TPJ), superior temporal sulcus (STS), and precuneus (PC).[9-12] There is evidence that activity in these brain regions is either dampened or enhanced by emotional reactivity. The amygdala is also connected to the brain's theory of mind network and plays a role in regulating our response to others.[13]

A group of women took a startle test in which loud white noise sounded or a sudden blast of air hit their face as they looked at disturbing photos. Then they played a competitive game while in an fMRI scanner, in which the winner could choose to punish the loser with a sudden blast of air. Women who had been highly reactive in the startle test showed the least activity in the temporal parietal junction (TPJ), medial prefrontal cortex (mPFC), and precuneus (PC) when deciding how to punish other women in the game. In other words, the areas of the brain responsible for empathy, in the most reactive women, became less active when threatened. In contrast, even after being provoked, women with low emotional reactivity had a richer theory of mind and a higher tolerance for being provoked.[14]

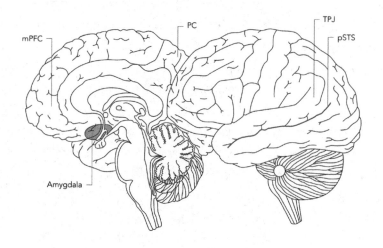

This link between temperament and theory of mind in humans means that, during our evolution, selection on emotional reactivity could have increased our tolerance as well as our ability to communicate cooperatively. Acting on the different ways people react to one another, natural selection might have been central in shaping our cultural cognition. This points to the possibility of human self-domestication.[15, 16, 17]

TAKE CONTROL

There was one problem with our human self-domestication hypothesis, as Richard began to call it.[17] We were suggesting that, as in other domesticated animals, the connection between emotional reactivity and theory of mind might explain our cognitive evolution. The problem was that our cognition, especially our theory of mind skills, outstrips that of other animals by leaps and bounds. If self-domestication was so crucial to our species' success, then why hadn't the same cognitive advances occurred in other self-domesticated species? Especially bonobos, who are so genetically similar to us? As Mike said, "Why aren't bonobos driving cars?"

It took me almost a decade to stumble on the answer.

Self-control is one of those cognitive abilities you do not appreciate until you lose it. Its source is the prefrontal cortex (PFC).[18] It is sometimes called the brain's executive center, because, like a good CEO, it stops you from making counterproductive or dangerous mistakes.

Self-control overrides the nucleus accumbens that tempts us to gamble; the visual cortex that sees a mirage in the desert; and the amygdala that makes us jump at noises in the dark. It is the space between a thought and an action, the look before the leap. Without self-control, we would all be divorced, in prison, or dead.

Some people have more self-control than others, and by studying these variations, researchers have demonstrated how central this trait is throughout our lives. One test of self-control is the famous marshmallow test, in which researchers give four- to six-year-old children a marshmallow and tell them they can either eat it right away or wait until the researcher returns and receive more marshmallows. Some children ate the marshmallow immediately, while others waited ten and even fifteen minutes without giving in to the temptation.[19]

Children who ate the marshmallow right away were more likely to struggle in school, have trouble paying attention, and have difficulty maintaining friendships. In various studies, when these same children grew up, they were more likely to be overweight, earn less money, and have criminal records.[20–22]

Self-control is just as important for nonhuman animals when they make decisions, and some species of animals seem to have more self-control than others. The biologist Evan MacLean and I came up with an easy way to compare levels of self-control between distantly related species— a kind of marshmallow test for animals.[23]

We put a treat inside a plastic cylinder that was open at both ends but fitted with a cloth that made it opaque. An animal could watch and remember as we placed the food inside. Then, after introducing this simple hiding game, we introduced our self-control test. We changed the cylinder in a way that at first would seem to make the problem easier. We simply removed the cloth and made the cylinder transparent. This meant an animal could see the food as they went to retrieve it.

We recruited more than fifty researchers from all over the world to use our cylinder test with over 550 animals from 36 different species, including birds, apes, monkeys, dogs, lemurs, and elephants.

All the species easily retrieved the food they had seen hidden a few moments before in the cylinder. But giving them more information, by making the tube transparent, actually made the problem harder. Now the correct solution required resisting the urge to grab the food directly through the transparent cylinder.

It sounds simple, and some species spontaneously solved it on their first go. But most species could not control their urge to reach. They went directly for the food and bumped into the solid cylinder even though they knew from their warm-up that they could only retrieve the food through the open ends. Some species, like the great apes, quickly learned to inhibit this reaching response after one or two mistakes, but other species, like squirrel monkeys, never learned— even after ten chances. We used the results to test big ideas about what leads some species to be more cognitively sophisticated than others.

I thought, as has long been argued, that animals that lived in larger groups—creating more complex social relationships— would require more inhibition to navigate life successfully. Instead, we found that the animals who passed their own marshmallow test simply had brains with more raw computing power. The small-brained animals we tested struggled with self-control, while larger-brained animals mastered the test almost immediately.[23]

The neuroscientist Suzana Herculano-Houzel has a theory that suggests why this might be. Herculano-Houzel was the first person to accurately count the number of neurons in animal brains. She would dissolve a brain into a thick, even soup and count the neurons in a sample of known volume. She found that as mammalian brains got bigger, they had more neurons in their cerebral cortex. This increased number of neurons might explain better self-control, but most mammals have to deal with a trade-off: As mammal brains get larger, their neurons expand and become less dense, as water dilutes a soup. For most mammals, there is a limit to how much computational power can increase with just an increase in brain size.

But primates break this rule. As primate brains get bigger, they grow more neurons, but even as their brains get bigger, their neurons stay the same size. In order to keep their neurons connected, primate brains have to pack in more neurons as they grow. As primate brains get bigger, their brain soup stays thick. For example, a capybara, a rodent the size of a pig, and a rhesus macaque, a monkey the size of a cat, have similar-sized brains, but the rhesus monkey has a brain with almost six times more neurons in its cerebral cortex, which likely means more computing power, and more self-control in the same sized brain.[24] Knowing this relationship between brain size, neuron density, and self-control suggests a very surprising and straightforward way that intelligence can increase—again, as a by-product. [24, 25]

Humans take the primate rule to the extreme.[26] Over the past two million years, human brains have essentially doubled in size, making them almost three times the size of a chimpanzee or bonobo brain. This has left us with higher neuron density in the brain's cortex than any other animal. It also explains the unprecedented levels of self-control we observe in our species. It might seem that as human levels of self-control increased, our other cognitive superpowers, including theory of mind, planning, reasoning, and language, simply came online and behavior and complex cultural traditions unique to our species followed.

The first problem with this scenario is that our brains were already within the size range of modern humans at least 200,000 years ago, but evidence of unique forms of modern human behavior do not appear extensively in the fossil record until around 50,000 years ago.[15] The second problem is that we were not the only hominins with big brains. As mentioned in the introduction, there were at least five other human species alive, and some of them had brains that are within the range of brain size we see in modern humans today.[27] All these large-brained humans had already evolved more than half a million years ago, and it is likely they all had self-control similar to or exceeding our own. And they all went extinct. Even at their height, their populations were sparse and their technology, though impressive, remained limited. Meanwhile, our own lineage did not develop explosive cultural complexity for more than 100,000 years after modern brain size and self-control appeared.

This relationship between brain size, neuronal densities, and self-control is what prompted me to think about extinct humans in a new way. Diet did not distinguish us from other humans: All hominins living within the past half million years likely controlled fire, cooked, ran long distances, and used tools to kill and butcher animals. We were not set apart by the size or density of our brain. Other humans, like Neanderthals, had culture within the range of ours, and perhaps even linguistic abilities comparable to our own. Nor, for thousands of years, was our technology better than anyone else's. This left one important difference between us and all the others: A little more than 50,000 years ago, we experienced a rapid expansion of our social networks.

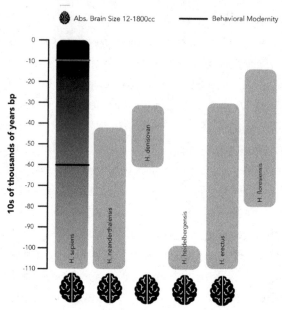

Self-domestication occurred at least by 80,000 years ago, before behavioral modernity.

Social networks are crucial in many ways, but they are essential to the development of technology. When populations lose their connection to a larger social network, technology does not just stop advancing—it can even disappear. Michael Tomasello has said that a child alone on a desert island would have culture very similar to a chimpanzee's.[28] The Tasmanian Aboriginals became isolated from mainland Australia around 12,000 years ago. Before this time, the fossil record shows that the nature and number of their tools were largely identical to those of the much larger population of Australian Aboriginals. But 10,000 years later, while the mainland Aboriginal toolkit had grown more impressive, the Tasmanian Aboriginal toolkit had withered to only a few dozen.[29]

Similarly, several hundred years ago, a population of Inuits settled in the Arctic Circle. When an epidemic winnowed the population down to a few hundred people, the community lost the ability to make kayaks, bows, and fishing spears. They were stranded and unable to hunt caribou or fish effectively. When they were discovered by another tribe of Inuits, they quickly regained the technology they had lost.[29]

The expansion of our social networks initiated a powerful feedback loop.[30] With expanding social ties, we could develop better technology. Better technology allowed us to hunt more food with less effort, which in turn allowed us to feed more people and live in denser groups. Denser groups lead to even better technology, and so on.[31–33]

But what set this feedback loop in motion? Population density can lead to innovation, but it can just as easily lead to violence as people compete for scarce resources. Even in the twenty-first century, with all our technology, rapid population growth can harm the environment and public health and increase rates of violence. What kept conflict in check while technology caught up to our needs? And why did it not appear in other big-brained, cultural humans? The human self-domestication hypothesis proposes that friendliness in the Pleistocene is the spark that ignited *Homo sapiens'* technological revolution.[34]

The human self-domestication hypothesis posits that natural selection acted on our species in favor of friendlier behavior that enhanced our ability to flexibly cooperate and communicate. Over generations, individuals with hormonal and developmental profiles that favor friendliness, and thus cooperative communication, were more successful.

This theory predicts we will find evidence for (a) selection for reduced emotional reactivity and heightened tolerance linked to new types of human cooperative-communicative abilities, and (b) changes in our morphology, physiology, and cognition resembling the domestication syndrome seen in other animals.

In the case of *Homo sapiens*, this selection acted on a *human* ancestor that was already a large-brained cultural being. Other animals might be self-domesticated, but only we were already in possession of extreme self-control at the start of the process. Our ability to carefully weigh our actions was further enhanced by reduced reactivity that occurred through self-domestication.

The human self-domestication hypothesis predicts that our expanded tolerance increases the reward for social inter-actions, as in bonobos and dogs. But it also predicts that we are unique because we can reliably inhibit our emotional reactions and intentionally calculate the benefits of toler-ance. It is this self-control combined with reduced reactivity that creates the human-specific adaptation for unique forms of social cognition.

Domesticating a wolf brain or an ape brain is impressive. But when you domesticate a *human* brain—this is when the real magic begins. An ultracultural species is born. A unique type of friendliness must have evolved in our species that allowed for larger group sizes, higher population densities, and more amicable relations between neighboring groups that in turn created larger social networks. This encouraged the transmission of more innovations between more innova-tors. Cultural ratcheting went from slow and sporadic to fast and furious. The result was exponential growth in technology and the emergence of behavioral modernity.

If the self-domestication hypothesis is correct, then we thrived not because we got smarter, but because we got friendlier. And though there is no human equivalent of Belyaev's fox experiments, we are lucky in one respect—domestication fossilizes. If self-domestication played a central role in driving the cultural revolution that we observe around 50,000 years ago, we should see fossil evidence before this time period. So it is just before this, at the mark of 80,000 years ago, that we began looking for evidence of self-domestication.[35]

DOMESTICATION IN OUR FACES

Selection for friendliness causes physical changes in domesticated animals. If humans are self-domesticated, there should be evidence for these types of physical changes in our ancestors. In the friendly foxes, selection based on behavior caused changes to their hormones during development. In turn, these hormones changed how the foxes grew.

In fact, there are hormones that regulate the development of appearance and behavior in humans. As you grow, both the length of your face and your brow ridge are modulated by testosterone. The more testosterone you have available during puberty, the thicker your brow ridge and the longer your face. Men tend to have thicker, more overhanging brow ridges and slightly longer faces than women,[36, 37] so we describe long faces with thick brow ridges as masculinized.

Testosterone has many different roles in your body, from jump-starting puberty to developing red blood cells. But it is best known for its relationship to aggression. Testosterone does not directly cause aggression in humans. People who are artificially given testosterone do not become more aggressive, even though this effect has been seen in some animals. But testosterone levels and their interactions with other hormones do seem to modulate aggressive responses, especially during competition.[38] The opposite effect is seen in men who are in long-term relationships or have newborn babies. Committed men and fathers show a reduction in testosterone that is thought to facilitate caretaking rather than competitive or aggressive behavior.[39]

There is evidence that women subconsciously judge men with more masculine faces to be more dishonest, uncooperative, and unfaithful,[40] and bad fathers.[41] In experiments, men also subconsciously estimate an opponent's strength just by looking at how masculinized his face is.[42] All of these findings help us read the faces of the past. And because there is a link between the development of behavior and physical appearance, we can look for physical changes in the fossil record to mark past behavioral changes.

Remember, we predicted that selection for friendliness must have begun having an effect by at least 80,000 years ago, before our population exploded and technology improved. To test this idea, we could compare human skulls from before and after this time. Juvenile behavior is friendly behavior. If our prediction was correct, we would see more juvenile-looking faces in the adults of our recent ancestors. These friendlier faces in the fossil record could be the signature of a people able to develop the sophisticated cooperative communication that allowed for the rapid growth in both our population and technology.[43]

To test our prediction, Steve Churchill and his student Bob Ceiri[34] analyzed the brow ridge projection and face shape of 1,421 skulls, including thirteen Middle Pleistocene skulls from 200,000 to 90,000 years ago and forty-one Late Pleistocene skulls from 38,000 to 10,000 years ago.[34] To measure facial width and height, they used the distance from cheek to cheek and from the top of the nose to the top of the teeth. To measure brow ridges, they examined how far the bone above the eyes projected from the face as well as its height above the eyes. The difference over time was dramatic.

The most visible change in the skulls was in the brow ridges. On average, skulls from the Late Pleistocene had a 40 percent reduction in how far their brow ridges projected from the face. Late Pleistocene faces were also 10 percent shorter and 5 percent narrower than the older Middle Pleistocene skulls. While this pattern varied, it continued so that the faces of modern hunter-gatherers and agriculturalists were even more juvenile in appearance than those of their Late Pleistocene ancestors.*

The signature of friendliness was found not just in fossil faces.[44] Several of the ancient skulls in our study were from human remains found in Es Skhul Cave in Israel. While we compared their brow ridges and facial length, the paleontologist Emma Nelson[45] measured their fingers. Like all primates, human mothers with high levels of androgens while they are pregnant have babies with a longer ring finger than index finger. This creates the low 2D:4D ratio we saw in chimpanzees

---

* The only exception to the pattern was a slight increase in the length of the face of the first farmers.

compared to bonobos. Men typically have a lower 2D:4D ratio than women, so we call a low ratio masculinized. In both people and animals, a more masculinized pattern of 2D:4D is associated with a greater degree of risk taking and potential for aggression.[45]

Nelson found that the 2D:4D of these Middle Pleistocene humans was lower, or more masculinized, than that of modern humans, suggesting higher androgens in utero. Nelson also showed that the 2D:4D of four Neanderthals was the most masculine of all. This suggests that our more feminized 2D:4D was not something we shared with other humans. It showed up late, around the same time as our feminized faces.

Another sign of domestication in animals is smaller brains. On average, domesticated animals have brains around 15 percent smaller than those of their wild relatives.[46] Smaller brains are housed in smaller skulls, so we should find smaller human skulls in more recent human fossils if we are self-domesticated.

When Steve and Bob compared fossil skull sizes, they found evidence that our skulls (and therefore absolute brain size) have been shrinking over the past 20,000 years—the period of our greatest intellectual accomplishment. Assuming similar body size, Steve and Bob found a 5 percent reduction in people's cranial capacity in the 10,000 years before agriculture, and then as agriculture took off.[34, 47]

In domesticated animals, serotonin is the most likely brain shrinker. An increase in the availability of serotonin is the first change we see as a domesticated animal becomes less aggressive.[48] There is also evidence that in mammals, serotonin is involved in skull development.

The effects of serotonin will be familiar to anyone who has taken Ecstasy. The active ingredient is MDMA (3, 4-methylenedioxymethamphetamine), which increases the availability of serotonin. MDMA floods the brain with up to 80 percent of the body's stored serotonin and prevents the reabsorption of serotonin into the brain. Users describe an overwhelming friendliness and the urge to hug everyone in sight.

Unfortunately, users also become familiar with serotonin deficit, because MDMA prevents the production of new serotonin. So dumping all the brain's serotonin on Saturday night usually leads to what some people call Suicide Tuesday. People who take Ecstasy report feeling more aggressive—and behave more aggressively in economic games—for a few days after using it.[49] Abnormalities in serotonin are also associated with violent criminal offenders, impulsive arsonists, and people with personality disorders.[50]

Selective serotonin reuptake inhibitors, or SSRIs, are antidepressants that prevent the reabsorption of serotonin in the brain, allowing more serotonin to be available, floating around the receptors. In experiments, people who take the SSRI Citalopram show an increase in cooperative behavior and are unwilling to harm others.[51, 52]

And this is where it gets interesting. Women taking Citalopram are more likely to have babies with smaller skulls.[53] Pregnant mice who are given Citalopram have babies with shorter, narrower snouts, and skulls described as globular.[54] Serotonin does not just change behavior. If there is more available serotonin early in development, it also appears to alter skull and face morphology.[55]

Not only did our skulls and brains shrink in comparison to those of other human species, our skulls have changed shape. Every other human species had a low, flat forehead and a thick skull. Neanderthals had heads shaped like footballs. *Homo erectus* had a head like a loaf of sandwich bread. Only we have the balloon-like skulls that anthropologists call globular.[56, 57] This shape indicates a possible increase in the availability of serotonin during our development. Like domesticated animals and Citalopram babies, our skulls shrunk, and like the Citalopram mice, our skulls grew rounder. Based on the fossil record, these changes started after we split from our common ancestor with Neanderthals.[58, 59]

So our faces, fingers, and skulls show signs of domestication, but what about the trademark of domestication—pigmentation? Over the generations, Belyaev's friendly foxes increasingly exhibited tawny reds, white stars on their foreheads, and other splotches of black and white. Many bonobos are missing pigment in their lips and tail tufts.

Apart from occasional abnormalities like piebaldism and vitiligo, people tend to have fairly uniform skin color. But in one part of our bodies, a change in pigmentation has made an enormous difference. Humans and domestic animals are the only ones to have pupils that display a variety of colors throughout life regardless of age or sex.[60] Our own colorful irises are visible because they are displayed on a unique white canvas, the sclera. Our sclerae are white because they are missing pigment.

Chimpanzees, bonobos, and all other primates produce

pigment that darkens their sclerae so that they blend in with their irises. This reduced contrast makes it difficult to see where or what they are looking at.

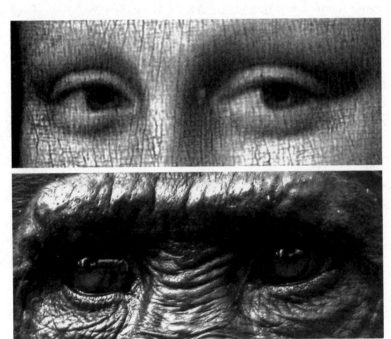

We are the only primate with white sclerae. Our eyes are also almond-shaped, making more of our sclera visible and allowing others to detect even subtle shifts in eye direction. At some point, we went from camouflaging our eyes to advertising them.[61]

From the moment we are born, we depend on eye contact.[62] We are born so much more helpless than other animals that even a few moments alone can be dangerous. To enlist the help we need to survive, we use our eyes. A baby's gaze triggers a release of oxytocin in its parents that makes them feel loving and loved. When parents look into the eyes of their babies, it triggers oxytocin in the babies as well, making them want to stare at their parents more.[63] Without it, our parents would not have been as motivated to get us through those first three months of life before we begin to laugh or smile.

Our eyes are also designed for cooperative communication. As babies, when we first recognized that our parents had intentions, feelings, and beliefs, we started paying attention to where they were looking and what they were pointing at.[64, 65] Our first meaningful experiences are built on these early cultural interactions.

The direction someone is looking can communicate their intention to play with a certain toy or move in a certain direction. Babies can then coordinate their actions in anticipation of playing together or being picked up. When toddlers learn words, they connect the sound an adult is making to the object the adult is looking at.[66] But not any eyes will do. Even at several weeks old, babies prefer eyes with white sclerae. They stare longer at cartoon eyes with white sclerae and dark pupils than at the reverse. Children prefer playing with stuffed animals that have white sclerae and dark pupils. Even adults prefer toys with white sclerae and dark pupils, though they are unaware of this preference.[67]

We are the only species that prefers white sclerae or relies on eye contact. Human babies can follow the direction of someone's gaze even when a person just moves their eyes. Chimpanzees and bonobos follow gaze direction only when a person moves their whole head, and they will even follow that direction when the person has their eyes closed. For all their understanding of what others can and cannot see, chimpanzees and bonobos do not seem to understand that sight depends on eyes.[68]

There are dedicated neurons in our brains that respond only when we see eyes. Located in the superior temporal sulcus (STS), these cells are part of the theory of mind network connected to subcortical emotional centers, including the amygdala. This neural network develops early.[69] By the age of four months, babies can already focus on the shape of the sclerae in someone else's eyes to understand their emotions. The neurons in this cortical network fire automatically and beneath our notice. Have you ever had the feeling while you were driving that someone in the next lane was watching you, and then you turned and someone was? That creepy feeling comes from the unconscious warning your STS sends the amygdala when your peripheral vision detects these eyes.[70, 71]

Most animals hide their sclerae to prevent their competitors from guessing what they might do next. But white sclerae gives human babies an advantage. The economist Terry Burnham and I reasoned that our unique eyes might also improve our ability to cooperate as adults. To test our idea, we organized a public goods game in which we gave people a certain amount of money and asked them how much of it they wanted to donate to a common fund. Good cooperators would donate more to the fund, while cheaters would keep most of their money.

As they decided, half the players were given the instructions by Kismet, a robot with giant white sclerae. The other players were given the instructions the exact same way but without seeing Kismet. Kismet had a startling effect. People donated around 30 percent more when he was watching.

This "Kismet effect" has been replicated outside the lab. People were less likely to leave trash in a public place if they were given a leaflet with eyes on it. Office workers were more likely to leave money for milk in the office breakroom if the donation box had eyes above it. Photos of angry eyes above bicycle racks reduced bicycle theft by 60 percent.

White sclerae seem to promote cooperation throughout our lives.[70, 72, 73] The self-domestication hypothesis predicts that we would have evolved white sclerae more than 80,000 years ago as a result of selection for friendliness. Increasing eye contact would have promoted the expression of oxytocin, encouraging bonding and cooperative communication. It would also discourage cheaters.[15]

Not only are our eyes unique and obvious, they are also universal. We have different-colored skin, hair, and even nails. Our irises can be green, gray, blue, brown, or black. But our sclerae are always white. It is unusual to have a trait that has absolutely no variability.

We use white sclerae to identify someone as human or human-like. Mickey Mouse became popular after his black spot eyes in *Steamboat Willy* were replaced with the giant white eyeballs in *The Sorcerer's Apprentice*.[74] And artists reconstructing extinct humans always give them white sclerae, even though there is no evidence for this in the fossil record. They seem to intuit that their models will *feel* more human if the models have eyes that are like ours. Interestingly, the fastest way to dehumanize someone is to paint their eyeballs black. In horror movies, white eyeballs are always the first to go. Even a slight change to the color of the eyeball is enough to make us uncomfortable, as when cute fluffy Mogwais with white sclerae become red-eyed gremlins.

If we are right, then only *Homo sapiens* has white sclerae as a result of self-domestication. Other humans, like Neanderthals, would have hidden their eyes with pigment, like every other primate. When we met these other humans for the first time, their darkened sclerae would have sent a strong signal that they were not like us.

5

Forever Young

We have seen the link between an increase in friendliness in humans and the accidental changes we believe it caused—including our feminized faces, white sclerae, and cognitive skills like cooperative communication. We know that an increase in friendliness triggers the self-domestication syndrome. But how do these changes actually happen?

The key is development. Changes in an animal's patterns of growth can be a powerful evolutionary engine.[1,2] Even a slight adjustment to the rate or timing of development can result in a completely different body type. For example, baby salamanders have gills, fins, and a tail—just like tadpoles. As they become adults, they lose their gills, replace fins with tails, and grow legs that allow them to walk on land. But one type of salamander, called the axolotl, keeps its gills and does not grow hind legs. Even as an adult, it still looks like a big version of a salamander that never quite grew up.[1]

Development can also affect social behavior. Young

cockroaches are extremely social. They hang out in groups, groom each other, and as a true sign of affection, eat each other's poop. It is not until they are adults that they become antisocial and solitary.

Baby cockroaches are wingless, with underdeveloped eyes and powerful gut bacteria that can digest anything—even wood—just like termites. In fact, termites are their closest relatives and are basically baby cockroaches frozen in their super-friendly juvenile state. They live in a colony where sterile workers serve a reproductive queen, and their ability to cooperate ensures their success.[3]

Appearing young can be advantageous. In one kind of crow, juveniles have a white spot on their beak that they lose when they become adults. Adult crows can be quite aggressive with one another. But when researchers painted a white spot on the beak of adult crows, other adults did not attack them. When the white spot was removed, they became fair game.[4]

It is not just looking young but acting young that can protect you from aggression. When young mice are afraid, they stand still and shiver. When other mice see this shivering, they usually pet or lick the young mouse to reassure them. Like Dmitry Belyaev, the psychologist Jean-Louis Gariépy bred mice based on their friendliness. But while Belyaev bred foxes based on their friendliness toward people, Gariépy bred his mice based on their friendliness toward other mice.

After six generations, Gariépy's friendly mice were far more tolerant than regular mice. Adult mice are usually aggressive toward strangers. But when the friendly mice saw a strange mouse, instead of being aggressive, they shivered like a juvenile mouse, and other adult mice were less likely to attack them. As with the foxes, selection for friendliness led to the retention of friendly juvenile behavior in adults and reduced aggression between mice.[5]

Even fish show evidence for the relationship between selection for friendliness, developmental change, and physical and cognitive change. Cleaner wrasse fish are small fish who clean parasites from larger fish. The larger client fish could easily eat the cleaner fish, but they never do. Observations of cleaner stations (where fish gather to be cleaned) show that as the cleaners clean, the normally predatory client fish become passive and nonaggressive (to both the cleaner fish and all other fish at the cleaning station).[6] A beautiful cooperative relationship has evolved in which one fish gets a meal and the other gets parasites removed.

As juveniles, all cleaner wrasse have distinctive mouths. As adults their mouths change shape and they find food in other ways. One species, the blue streak wrasse, maintains its juvenile mouth into adulthood and continues to get food by cleaning other fish.[7] Like dogs, these fish have evolved as specialists in friendly interaction with other species. This life of cleaning also effects their cognition. In experiments, blue streak wrasse are better at cooperating than other closely related wrasse who do not clean as adults. This allows blue streak wrasse to resist biting their clients, which would be more nutritious, and focus on eating their parasites instead.[8, 9] Changes in serotonin and oxytocin levels consistent with extended juvenile development modulate the friendly behavior of these cleaning fish, just as they do in experimentally domesticated species.[10, 11]

Dogs and bonobos not only maintain juvenile behaviors throughout their lives, they, like the wrasse, develop behaviors related to cooperative communication earlier than their relatives.

Shortly after a puppy's eyes open, it is ready to bond with other dogs and people.[12] In this same period, it is motivated to explore new places and things.[13] More time to develop allows more time to gain a variety of experience. Experience gives dogs the confidence to deal with a constant stream of new people, places, and things that would overwhelm a wolf in the city.

The window for socialization also stays open longer. While this hyperexploratory period lasts for weeks in wolves, it lasts months, if not years, in dogs—who maintain or even increase puppy-like responses to novelty as they mature.[14]

Even the way dogs vocalize has been affected by this expanded window of development. While both dog and wolf puppies bark to get the attention of their mothers, only adult dogs continue barking so frequently and in so many different contexts.[15]

We know that the expanded window of development in dogs that allows for these cooperative behaviors is based on a selection for friendliness because Belyaev's foxes show the same developmental patterns.[16] The regular foxes, like wolves, have only a short window—from sixteen days to six weeks old—in which they can be socialized to people. Like dogs, the friendly foxes have an expanded window of socialization that opens earlier, at day 14, and closes later, at week 10.[17] And just like dogs, the friendly foxes maintain puppy-like vocalizations throughout their lives. Even when fully adult, they bark and whine like fox kits at the sight of a human, while the regular foxes do not.

In bonobos, there is one type of social behavior that appears very early. At lunchtime at the bonobo sanctuary in Congo, someone brings a giant fruit salad to the nursery where all the baby bonobos play. The basket is piled high with mangoes, bananas, papayas, and sugar cane.

As soon as the baby bonobos see the food, they start to peep. As they become more excited, they grab the bonobo closest to them and rub their genitals together. There is no actual penetration; it is more a wild kind of frottage. These babies are orphans, separated too young from their mothers and other bonobo adults to have learned this behavior. We did not see this behavior in the chimpanzee sanctuary. While bonobos show sexual behavior as infants, in chimpanzees this behavior does not appear until around puberty.[18]

There is some evidence that the mechanism behind this sexual behavior is hormonal. Infant bonobos already have the testosterone levels of juvenile chimpanzees. Bonobos maintain these juvenile levels of testosterone into adulthood.[19, 20]

These levels may be tied to faster reproduction in bonobos. While most bonobo sex is nonreproductive, female bonobos are, just like domesticated animals, able to have babies at an earlier age than their more aggressive relatives.[21] But bonobos use their sexual behavior throughout their entire lives to break up fights, calm young bonobos who are upset, and strengthen their friendships with other females.[18]

Selection for friendliness is really selection for an expanded window of social development. In dogs and bonobos, this means that key traits for social flexibility develop earlier and continue growing later.

UNDER THE HOOD OF DEVELOPMENT

How exactly do genes for development evolve to cause the self-domestication syndrome? Some genes have a much bigger role in determining how an organism develops than other genes. These genes can control how hundreds of other genes end up doing their jobs.

This is not what I was taught in high school genetics. I learned about Mendel's pea plants, with their dominant and recessive genes, dominant traits being those most usually expressed. Mendel found that different traits are inherited separately and do not depend on each other. Each gene has a different job, like controlling the color or shape of the pea flowers. This independence creates the opportunity for diversity in form and function, since pea plants can produce different color or shape combinations. Selection then acts on this variation.

But decades of research have shown that development makes this picture more complicated. Mendel's genes reveal just one of the simplest ways heritable variation is created. While different genes can have different jobs, one gene, or a set of genes, can have multiple jobs. For example, a gene can be involved in both bone growth and pigmentation. These multitasking genes may do two jobs simultaneously or at different times during your life.

There is another class of genes that are like librarians. Genes are like books filled with instructions on how to make different proteins. These proteins are the building blocks of every fluid and tissue in our bodies—including our brains. Librarian genes in each cell of our bodies make recommendations about which books should be read and how often, and some librarian genes have control over large parts of our genetic library. Librarian genes can have a big impact, since they control not only which protein is made, but the rate and time at which it is made.

A tiny change to one of these multitasking or librarian genes can have a huge impact on many traits at the same time. This is especially true for multitasking and librarian genes that help control development. The earlier they kick in and the longer they act, the more they will amplify genetic change. This is how small genetic changes that help control development can alter an animal so dramatically that it becomes a different animal entirely, just as we saw in termites, axolotl, and wrasse.[2, 22] A similar explanation has been proposed for the changes we see in dogs, bonobos, and even our own species.

THE NEURAL CREST

The discovery that traits can emerge as by-products of friendliness is one of the most remarkable of the last century. Belyaev and his team selected the foxes only to be friendlier toward humans, discounting any other cognitive, physiological, or morphological characteristic. Yet selecting foxes for friendly behavior led to foxes with a higher frequency of shorter, curlier tails; piebald, multicolored coats; shorter snouts with smaller teeth; floppy ears; a longer reproductive period each year; high levels of serotonin; and enhanced cooperative communicative abilities. The changes seen in the foxes are even more dramatic given that they mirror those commonly seen in other domesticated mammals—as well as species we suspect are self-domesticated.

In attempting to explain the link between friendliness and the suite of traits that tend to make up the mammalian domestication syndrome, Richard Wrangham and the geneticist Adam Wilkinson became particularly interested in neural crest cells, which play an outsized role in development.[23] Neural crest cells appear for a short time in all vertebrate embryos. These cells develop on the back of the neural tube, which eventually becomes the brain and spine. Neural crest cells are stem cells, which means that as the embryo develops, neural crest cells can transform into a variety of cell types. They are also migratory. As the embryo develops, these special stem cells travel all over the body. A set of librarian genes is thought to have powerful sway on which type of cell these stem cells become and when and where they migrate.

Migratory neural crest cells play a role in the development of a host of traits associated with the domestication syndrome. Central to domestication is reduced fear and aggression; neural crest cells are involved in the development of the adrenal medullae that release adrenaline.[23] In domesticated animals, the adrenal glands are smaller than in their wild cousins, and smaller adrenal glands mean fewer stress-inducing hormones. Neural crest cells are also heavily involved in all of the tissues altered in association with the selection for friendliness. This includes the development of tail and ear cartilage, skin pigment, the bones of the snout (or face), and the teeth.

The neural crest cells in the head are also thought to affect brain development.[24] This may account for changes in brain size as well as for how receptive different areas of the brain are to neurohormones like serotonin and oxytocin.[25] These brain changes are likely linked to changes in reproductive cycles as well. A smaller brain size could affect the hypothalamic pituitary gonadal (HPG) axis that controls reproductive cycles. Limiting the function of the HPG axis would produce earlier sexual maturation and more frequent breeding cycles.

Wrangham and Wilkins predicted that in the domesticated mammals (and perhaps birds), the initial selection for friendliness would have favored librarian genes that changed the way neural crest cells develop. This would have caused, almost to the letter, the changes we see as part of the domes-

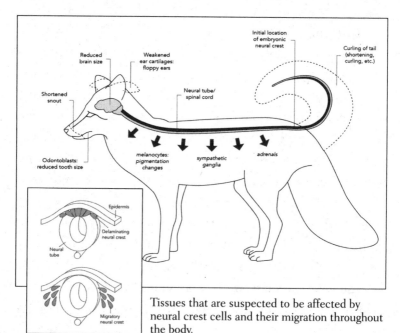

Tissues that are suspected to be affected by neural crest cells and their migration throughout the body.

tication syndrome in every species from dogs to bonobos. Changes to library genes controlling neural crest cell development and migration provide a powerful explanation for the seemingly improbable link between friendliness and the other attributes associated with domestication.[26] Constantina Theofanopoulou and Cedric Boeckx, neurobiologists at the University of Barcelona, have examined whether this link might exist in humans, using ancient DNA. They discovered that the same genes that evolved in many domesticated animals have also evolved in humans.[27] This includes neural crest genes that have been modified since we split with other extinct human species.[28] This provides the first genetic evidence for human self-domestication.

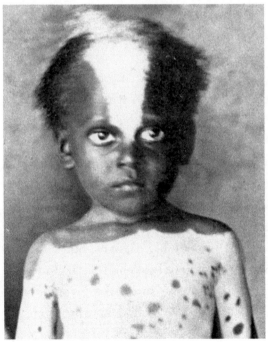

Lisbey, five years old in 1913, displays neurocristopathy, a condition in humans that is caused by altered neural crest cell development. A symptom of neurocristopathy is piebaldism, a condition characterized by coloring remarkably like what we see in domesticated animals.

EARLY TO GROW, LATE TO FINISH

Serotonin changes the shape of our skulls. Androgens like testosterone alter our faces and hands. Loss of pigmentation in our eyes dramatically enhances our cooperative communication. All of these changes point to selection for friendliness late in hominin evolution.

The progress of our development is a big part of what has made us different from other, extinct humans.[29] Compared to them, and to other primates, we have an unusual life trajectory. We are born too early and we take forever to reproduce; but then we are able to reproduce faster with a shorter interval between babies, and women go through menopause with decades left to live.[30] Our cognition also marks us as different—both its early expression and the extended development of traits related to cooperative communication and tolerance.

We are born with a brain only a quarter the size of our adult brain, while other apes are born with a brain almost half its adult size. This means that as babies we are exceptionally helpless.[31] Then, before we can run or climb, between nine and twelve months, we start thinking about the minds of others, first in simple, then in increasingly complex ways.

We suspected that this pattern of development was unique to humans. We had not seen these same cooperative communicative skills in other adult great apes, but without a direct comparison of development between human and great ape infants, we could not know for sure. So we worked with bonobo, chimpanzee, and human infants, starting from the age of two. We tested almost a hundred infants from the three species on almost two dozen cognitive measures every year for three years, to see how their cognition developed. We measured every type of cognition that made sense for their age: counting, causality, tool use, self-control, emotional reactivity, imitation, gestures, gaze following, and about a dozen more. It was the first time anyone had directly compared so many infants from all three species across so many different abilities. And we found that two-year-old human babies were unremarkable when it came to the nonsocial tasks, like counting and understanding the physical world. They looked just like bonobos and chimpanzees of the same age.

The difference appeared when we looked at how human babies solved social problems. At two years old, human babies, with their completely underdeveloped brains, showed more social skill than nonhuman apes with more mature brains. By four, human babies outstripped the other ape babies on every task. The same toddlers who can't put their drinks down without spilling them, or make it to the toilet in time, have a theory about how other minds work.[32]

These early-emerging social skills give us an edge, allowing even those of us with underdeveloped brains to use others to solve sophisticated problems. Our ability to understand others at a young age also allows us to inherit knowledge acquired over generations, giving us a unique survival advantage.

BABY BALLOONS

In order to mark the moment that we domesticated ourselves, we need to identify when we developed this unparalleled social intelligence. What makes this possible are fossilized *Homo sapiens* skulls that hint at brain development.

There are two relevant physical markers for our brain development. The first is the giant hole we have in our head when we are born. Unlike most mammals who are born with fully developed skulls, *Homo sapiens* and Neanderthal babies' skull bones are unconnected so they can squeeze through the birth canal. Our shape-shifting baby skulls and the underdeveloped brains they contain evolved late, but not uniquely in our species.[31]

The second morphological signature of development is in the skull's shape. Unlike other humans, who had low, flat foreheads and thick skulls, our infants developed balloon-shaped heads to fit their oddly shaped brains.[33]

Other animals shut down brain growth soon after they are born, but we maintain fetal brain growth rates for two years past birth.[34] This rapid postnatal growth particularly affects the top and back of our heads, creating its balloon shape.[31] The top and back of the brain, the parietal regions, house two central nodes of our brain's theory of mind network—the temporal parietal junction (TPJ) and precuneus (PC) regions.[35] These are the areas that become active as babies start paying attention to the gaze and gestures of others.[36] We can infer from fossilized skulls that these early-emerging social skills were unique to *Homo sapiens*.

But the type of social cognition that develops early is very specific. Beyond cooperative communicative skills, other skills seem to be delayed. It is not until four to six years of age that human self-control begins to outstrip that of other apes, just in time for them to take the marshmallow test.[37, 38] Self-control develops so slowly that we do not have fully adult levels until our early twenties. This may be why, as teenagers, we take more risks—which is why car insurance for a sixteen-year-old is more expensive than it is for a twenty-one-year-old. Fortunately, teenagers also feel failure more deeply, so we are quick learners.[39]

This progression is a manifestation of synaptic pruning. When our brains are growing, we make more neurons than we need. As we navigate our lives, solving problems and adapting to different environments, we use certain networks of neurons more than others. These commonly used networks become more numerous and better able to compute information; then they streamline their connections and become more efficient. By the time we are adults, our brain networks are stripped down and specialized. We lose plasticity, but our cognition becomes better at solving the problems we are most likely to face.[39]

Other animals are born with brains that develop quickly in closed skulls. Baby wildebeest find their feet within minutes and can keep up with the herd within days. Even baby chimpanzees are up and about long before we are.

We are born, as the poet William Blake wrote, "helpless, naked, piping loud," and we stay that way for years. But because of our early-emerging social cognition, we can plug into other minds. We can read the intentions, beliefs, and emotions of others in a way that allows us to harness the efforts and love of our caregivers who keep us safe while our brains continue to form and flourish.

When we're babies, our minds have the power to read the intentions, beliefs, and emotions of others, and we use this power to compensate for our physical helplessness—our weak muscles and unformed skulls—as our brains slowly catch up, growing, then pruning, so that when they are finished in our early twenties, they have become supercomputers, uniquely designed to learn and innovate in the cultural environment into which we are born.

THE FRIENDS YOU HAVEN'T MET

We have seen dogs and foxes selected to become friendlier toward people and bonobo males who were selected to be friendlier toward females. What type of friendliness drove human self-domestication?

Each day, hunter-gatherers like the Hadza in Tanzania search for food and then return to camp to cook, eat, socialize, and sleep. Women share the tubers they have dug up and the fruits they have gathered. Men return with highly prized meat and honey.[30] Apes sometimes share food while they are foraging, but only humans bring food home.

In hunter-gatherer communities, food is distributed relatively equally.[40] The families of the most productive foragers do not get the lion's share of food. They share, and in return, they make friends who will take care of them if they are hungry, injured, or sick. With no crops, refrigerators, banks, or governments, these social bonds are their only insurance.[30]

The most productive Hadza hunters have already eaten enough to meet their daily calorie needs before they return to camp with food.[41] Sharing their surplus is mutually beneficial, since they feed others while reinforcing bonds that will buffer them against shortfall.[42] The prospect of sharing also motivates cooperation, since it means more food for everyone. Over hundreds of generations, these strongly bonded groups have developed a competitive advantage over less cooperative, more despotic humans with less reliable social insurance.

This insurance alters the calculus of social relationships. While chimpanzee cooperation is constrained by fear and despotism, hunter-gatherer cooperation rewards everyone. Unlike chimpanzee males working to dominate one another, hunter-gatherers use aggression to prevent individuals from dominating the group. When aggression within human groups is used to prevent power grabs rather than assert dominance,[30, 43] sharing, tolerance, and cooperation increase exponentially.

This new social formula meant that the benefit of new social partners was often far greater than the cost. This includes new social partners from outside the group. A similar calculation explains why bonobos are attracted to strangers. Without the risk of lethal aggression from neighboring males, female bonobos can interact with neighbors and expand their social networks.

Selection for friendliness occurred in us, but the difference is that we did not just become more tolerant overall, like bonobos. We expanded our definition of who we considered a group member. Chimpanzees and bonobos recognize strangers based on familiarity. Someone who lives with them, inside their territory, is a group member. Everyone else is a stranger. Chimpanzees may hear or see their neighbors, but the interaction is almost always brief and hostile. Bonobos are much friendlier with the unfamiliar.

We, too, respond to those we are familiar and unfamiliar

with in different ways, but unlike any other animal, we also have the ability to instantly recognize whether a stranger belongs to our group. Human group members are not defined geographically but have a broader kind of identity, unlike bonobos or chimpanzees.

A new social category evolved for humans—the intra-group stranger. We can recognize people we have never met as belonging to our group. Someone who wears the same sports jersey, fraternity tie, or religious symbol on a necklace. Each day, without thinking about it, we adorn ourselves in ways that make us identifiable to the other members. We are even prepared to care for, bond with, and sacrifice ourselves for intragroup strangers.

This capacity dominates our modern lives. We are sur-rounded by people we do not know, and yet we are not just tolerant of these strangers, we actively help one another. This friendliness encourages us to perform acts of kindness both great, like donating an organ, or small, like helping someone cross the street.

People do prefer to help strangers with whom they share a group identity—especially if they know that the stranger is aware of the bond.[44] The Tsimane of Bolivia are hunter-gatherers. When researchers showed the Tsimane pictures of strangers from their group and strangers from other groups, they were willing to share more with the Tsimane strangers.[45] Likewise, people who live in fourteen out of fifteen industrial economies shared more with strangers from their own coun-try than with those who did not identify themselves with the same nationality.[46]

We are babies when we begin to recognize intragroup

strangers—around the same time that our earliest theory of mind abilities come online. Nine-month-old babies prefer puppets who like the same food they do, and they prefer people who are nice to these puppets.[47] Seven-month-old babies prefer music introduced by someone in their native tongue.[48] Just as babies begin to pay attention to the communicative intentions of others, they also begin to choose where to focus their attention. Even very early in life, our thoughts, feelings, and beliefs about the thoughts, feelings, and beliefs of strangers are stronger for some strangers than for others. The psychologist Niam McLoughlin showed first graders a picture of a doll face morphing into a human face. The first graders had to let researchers know when they thought the picture "had a mind." The children were quicker to say an image had a mind when they were told the picture was of someone from their town than when they were told that the picture was of someone from far away.[49, 50] Children are also more generous toward others they believe share their group identity.[51] As adults, we are also more likely to acknowledge that others have a mind like ours if we believe they are from our group.[52]

Even though we are essentially born attracted to people who share our group identity, what constitutes that identity is highly influenced by social forces. Even for babies, group identity is about more than just familiarity. As we grow, it can be defined by almost anything: clothing, food preferences, rituals, physical traits, political affiliation, place of origin, or loyalty to sports teams. While we appear biologically prepared to recognize group identities, our social awareness allows our construction of these identities to be flexible.

This plasticity is what the anthropologist Joseph Henrich

argues is critical to the emergence of social norms.[53] Norms are the implicit or explicit rules that govern even the smallest social interaction. They are central to the success of all our institutions, and they must have arisen after we humans had domesticated ourselves, allowing us to identify and embrace humans beyond our immediate families.

GROUPS THAT FEEL LIKE FAMILY

The molecule largely responsible for the evolution of this new social category is probably the neurohormone oxytocin.[54] Oxytocin is closely linked to serotonin and testosterone availability, two hormones we have inferred were altered as a result of human self-domestication. More available serotonin impacts oxytocin, since serotonin neurons and the activity of their receptors mediate the effects of oxytocin. Basically, serotonin increases the impact of oxytocin. Decreases in available testosterone also increase the ability of oxytocin to bind with neurons and change behavior.[29] With increased available serotonin and decreased testosterone, during human self-domestication, an increase in the efficacy of oxytocin is predicted. It is this increased power of oxytocin over our behavior that likely explains how our species evolved the ability to perceive our group as if they are family.

When researchers gave people oxytocin to inhale, the subjects were more likely to empathize and could more accurately recognize others' emotions. The pathway for the hormone seems to be through the brain's medial prefrontal cortex (mPFC), part of the brain's theory of mind network.[55] Oxytocin may disrupt the connection between the mPFC and the amygdala, allowing the mPFC to have more influence and blunting the amygdala's fear and disgust response. In other words, oxytocin reduces feelings of threat and enables trust. When researchers gave people oxytocin to inhale, they tended to be cooperative, to donate more generously, and to be more trusting in financial and social games.[56]

Oxytocin floods through mothers during childbirth; it facilitates milk production and is passed on through breast milk. Eye contact between parents and babies creates an oxytocin loop, making both parent and baby feel loving and loved. The whites of our eyes, both highly visible and universal, help to jump-start this oxytocin loop. Though it probably originated to encourage caregiving between parents and their babies, it acts to encourage bonding in face-to-face interactions among all humans. Even dogs, but not wolves, can hijack this bonding pathway, creating an oxytocin loop between themselves and their people.[57]

According to the human self-domestication hypothesis, when we see an intragroup stranger, oxytocin should help us feel friendly toward them, like bonobos, rather than aggressive, like chimpanzees.[58, 59] Eye contact creates a further surge in oxytocin, reinforcing an emotional bond. When you meet someone new, your willingness to make eye contact long enough for the oxytocin to kick in is probably more important than the firmness of your handshake. A knack for befriending intragroup strangers improves our fitness, because, as we saw with the Arctic Inuits and Tasmanians, people in isolation lose cultural knowledge. Human cultural innovation was supercharged because hundreds and then millions of innovators were uniquely able to accept and cooperate with strangers.[60]

We think this new category of intragroup stranger appeared in our species during the Middle Paleolithic, more than 80,000 years ago, and allowed communities to become larger and more densely populated. The anthropologist Kim Hill has suggested that this level of tolerance was facilitated as both men and women immigrated into neighboring communities, binding families together across groups in a way not observed in other primates.[61]

As populations became more dense, technological innovation would have exploded. With improved technology, we could expand into a wider range of ecologies than other human species. Trade across networks of neighboring groups that shared rituals or a communicative system allowed us to transfer innovations far and wide. Strange ideas met even when their inventors did not. People could barter for far-flung natural resources and gain access to water across territorial boundaries.

This newfound coordination would have created the potential for collective action such as a multigroup hunt for large mammals or fish. Once the social calculus shifted in favor of friendliness, the increasing network of minds would have given us a major advantage over other human species.

THE NICEST HUMAN WON

The idea that friendliness led to our success is not new. Neither is the idea that as a species, we became more intelligent. Our discovery lies in the relationship between the two ideas: It was an increase in social tolerance that led to cognitive changes, especially those related to communication and cooperation.

Like the foxes under the watchful selection of a Russian genius, or the Bengalese finches singing loops and harmonies around their wild cousins, we have domesticated minds. And like those first bonobos settling into their rainforest below the arc of the Congo River, or those protodogs scavenging on our trash, we effected this change ourselves.

Domesticating a human, however, is not the same as domesticating a bird, a wolf, or even a great ape. Only humans have gigantic brains, packed densely with neurons, which, along with other cognitive abilities, gave us unprecedented self-control. We shared this self-control with other human species who were able to build rudimentary tools. Closer relatives, like the Neanderthals, had sophisticated culture, weapons, and perhaps even language, but they were never a top predator. They were on the level of a jackal or a hyena in the carnivore hierarchy, sometimes hunting, sometimes scavenging, and often at the mercy of larger carnivores.

Our species was not doing much better. Megadroughts, erupting volcanoes, and advancing glaciers threatened our survival, and we may have neared extinction. Then, in the Middle Paleolithic, we, and only we, underwent intense selection for friendliness.

This selection for friendliness gave us a new social category, one that no other animal has—the intragroup stranger. This category was sparked and maintained by oxytocin, the same hormone that floods through a mother at the birth of her baby. With oxytocin, even at a distance, we could feel kindness toward an approaching stranger we could see was like us.[50] Perhaps they had a certain ocher pattern on their body. Or a necklace with shells from our coastline. When they were close enough to grasp our hands, eye contact would have sent another pulse of oxytocin through us both. We would be less fearful, more trusting and cooperative.

Coupled with the extraordinary levels of self-control we shared with all humans, we could weigh the benefits of cooperation. We were better able to think through the consequences of our actions.

Human self-domestication before 80,000 years ago led to both the increase in population and the revolution in technology we see in the fossil record. Friendliness drove this technological revolution by linking groups of innovators together in a way other hominin species never could. Self-domestication gave us a superpower, and in the blink of an evolutionary eye, we took over the world.

And one by one, every other human species went extinct.

Our capacity for friendliness toward strangers has continued to increase. The psychologist Steven Pinker argues that human violence has steadily declined over time.[62] Yuval Harari writes that the "Law of the Jungle has finally been broken, if not rescinded . . . a growing segment of humankind has come to see war as simply inconceivable."[63]

We have human self-domestication to thank for this. The concept of the intragroup stranger has allowed us to extend our love toward those we have never met. This notion of an extended family helped us to succeed in the past, and it is the great hope for our future. As our population grows and uses more resources, we must keep expanding our circle of trust in order to flourish.

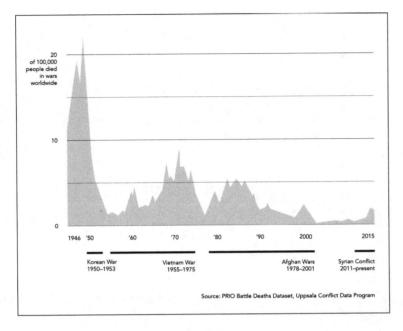

Source: PRIO Battle Deaths Dataset, Uppsala Conflict Data Program

And yet this optimistic view of ourselves is at odds with the misery and suffering we still inflict on one another.[64] Our theory of mind gives us a unique capacity for compassion, but at times it seems to go missing.

We know how human self-domestication explains the best in us, but does it also explain the worst? How do we reconcile our unique friendliness with our capacity for cruelty?

6

Not Quite Human

Our nanny, Rachel, sang as she gently wiggled a shoe onto our daughter's foot. Our daughter was clapping and bouncing when she knocked Rachel's skirt to one side, revealing a scar that went from her knee to the bottom of her shin.

"Rachel, what happened to your leg?" I asked.

The wound had not been stitched well. The tissue was swollen and crooked. Rachel looked at the floor and shrugged.

"Machete."

She pulled her skirt back to reveal an identical scar on the other leg.

Rachel was born in Minembwe, a small village high in the mountains overlooking Lake Tanganyika. She grew up like any other girl, going to school, playing in creeks, and running around the neighborhood with her friends. Her parents owned a store, and since Rachel could add long columns of numbers in her head, she worked the register after school and on the holidays and sneaked candy to her friends when she could.

Rachel is part of the Banyamulenge tribe from eastern Congo. Banyamulenge trace their lineage to the Tutsi in Rwanda and all the way back to the Queen of Sheba in Ethiopia. In the sixteenth century, they wandered over the volcanic mountains from Rwanda in search of grazing land for their cattle. They settled on the Ruizi Plains, at an altitude of

3,000 feet. The air was cool and wet, there were no tsetse flies, and the mountains were covered in thick grass.

As Rachel grew older, she realized that as a Banyamulenge, the world was not open to her in the same way it was to some of her friends. The Banyamulenge are known as the Black Jews of Africa. They are seen as immigrants, even though they arrived in Congo four hundred years ago and were only one of the many migrations in and out of the region. Rachel could not go to university. She was not allowed to live in Uvira, the closest city. She could never be a politician or become part of the local government. Occasionally, when she ventured outside her neighborhood, someone would mutter "Dirty Rwandan" as she passed.

In the nineteenth and then early twentieth century, while the Belgians were enslaving millions of Congolese to work in the rubber plantations, they took one look at the sleek Banyamulenge cows and knew the Ruizi Plains were worth their weight in gold. The Belgians imposed a crippling tax for each head of cattle. The Banyamulenge refused to pay, as the Belgians knew they would, and the Belgians drove them from their land.

On the other side of the border in Rwanda, the Belgians had elevated the status of the Tutsi, the tribe the Banyamulenge had left so long ago. It was Rachel's Tutsi heritage that gave her a thin nose and high neck. Even before the Europeans arrived, the Tutsis had higher status in society than the Hutu, who were described as darker-skinned, with rounder faces and flatter noses. But before the Belgians, there was a certain fluidity between the two tribes. There were mixed marriages, and a Tutsi could become a Hutu and vice versa.

The Belgians came to Rwanda with instruments to measure facial features and decided that the Tutsi more closely resembled Europeans and so were the superior tribe. Everyone was issued identity cards. Tutsis were given better positions, and easier access to resources and education. Hutus were relegated to poorly paid labor. The divide between the two grew deep enough to incite several massacres between them, culminating in the devastating Rwandan genocide of 1994.

The Belgians abruptly abandoned Congo to its independence in 1960, evacuating their citizens and letting the country descend into chaos. The population splintered into dozens of rebel groups. In the following decades, the cows of the Banyamulenge were a tempting target for hungry soldiers and the Banyamulenge themselves were attacked repeatedly, until young Banyamulenge men joined rebel groups to protect their homes and families.

The Banyamulenge never got a foothold in the scramble for Congo precipitated by the Belgian exit. Other tribes began to mutter that the Banyamulenge were not even Congolese. By the time Rachel was twenty-three, she had been stripped of her citizenship and barred from traveling and was no longer allowed to vote.

Despite these disadvantages, protected by her community in her mountain home above the clouds, Rachel was happy. She married a man she loved and they had two little girls who grew up running wild along the mountain paths, just as she had.

Rachel and her family survived the Congo War of 1996. They were battered but hopeful. Because Banyamulenge soldiers helped topple the former dictator and put the next president in power, Banyamulenge throughout the country thought they might finally be accepted. But the new president quickly turned against them, and when the next war came, in 1998, it was far worse than the ones before.

Rebel groups swarmed through eastern Congo, shooting people on sight and raping women in the fields. Rachel fled with her family to a refugee camp in Burundi.

The Gatumba refugee camp was mercifully close—only fifteen miles away. Rachel's parents, her brother and his family, her cousins, and most of her neighbors had left their homes with only what they could carry. They moved into tents. There were showers and toilets. They shared food and pots and clothes. The children played together on the fields. The adults told one another that it would be over soon and wondered whether it was safe enough for someone to go back for supplies. If they walked a few miles through the marsh to a fishing village on the shore of Lake Tanganyika, they could see home.

The United Nations Refugee Agency (UNHCR) was unhappy that the refugees at Gatumba were so close to the Congolese border and wanted to move them farther inland to camps with other Banyamulenge. Rachel and the other refugees refused to go. They wanted to be close to home, and they were afraid of rebels and the crowded conditions of the inland camp, which they heard was rife with disease.[1]

The UNHCR warned that the camp was closing down. They paid for ten policemen to guard the camp, but they removed the camp administrator and stopped supplying food. Still, Rachel and her family refused to move.

Then it was too late.

The rebel group that attacked their camp was called PALIPEHUTU, which stands for *Parti pour la libération du peuple hutu,* or the Party for the Liberation of the Hutu People. PALIPEHUTU was the only rebel group who had not signed a peace treaty with the Burundian government, despite being crushed by the Burundian, mostly Tutsi, army. There were only fifteen hundred rebels left, and some of them were children so young that their rifle butts dragged on the ground. Humiliated by their losses, the rebels looked around for something to smash.

On August 13, 2004, Rachel, her husband, and their two daughters were asleep when the rebels sneaked into their camp. She woke up to the sound of screaming and the smell of smoke. Through the confusion, she heard people singing "Hallelujah," and drums, bells, and whistles. Another clear voice sang above the rest, "God will show us how to get to you and where to find you."

The men cut open the tent with machetes. They killed Rachel's husband and her two daughters in front of her and dragged Rachel out of the tent. The sky was red. Almost every tent was on fire.

The 258 dead and wounded were all Banyamulenge. The rebels had lists with names and tent numbers. They stationed people outside the tents housing those from other tribes, warning them not to come out. A hundred Burundian soldiers and dozens of police were stationed close enough to hear the screams. They did nothing. The UN peacekeeping force got word only when it was over. The next morning, UNHCR officials wandered dumbfounded among smoldering earth and scorched corpses. Later, they would call it the Gatumba Massacre.

The rebels dragged Rachel into the jungle and raped her for a year. At some point, one of the men cut her legs with a machete, either to stop her from escaping or for no reason at all. One day, when most of the men were out, Rachel escaped. She made her way to a refugee camp in Zambia, more than a thousand miles from her home. When she arrived, she was barely alive. In the camp, a dirty needle gave her an infection that almost killed her. She spent four months not knowing whether she would live or die. By some miracle, the rebels had not gotten her pregnant. They had, however, given her HIV.

Rachel (right) at the refugee camp in Zambia. At the camp, Rachel helped other women with HIV, explaining what the disease meant and how to take their medication. She immigrated to America in 2009.

* * *

When groups of people feel threatened by one another, it opens up the dark side in both groups. The more powerful group might attack, as the Hutu attacked the Banyamulenge, or the attacked group might retaliate. Self-domestication points to the origin of our worst forms of aggression.

Dogs and bonobos are friendlier as a result of self-domestication, but both species have also evolved new forms of aggression toward those who threaten their families. Dogs aggressively bark at strangers near their human homes. In bonobos, their protective maternal style and the bond between females leads them to be more aggressive toward males than chimpanzee females are. We suspect these increases in aggression appear during self-domestication as a result of changes in the oxytocin system.[2]

Because oxytocin seems crucial to parental behavior, it is sometimes called the hug hormone. But I prefer to call it the momma bear hormone. The same oxytocin that floods through a mother with the arrival of her newborn[3] feeds the rage she feels when someone threatens that baby. For example, hamster mothers given extra oxytocin are more likely to attack and bite a threatening male.[4] Oxytocin is also implicated in related forms of male aggression. Available oxytocin increases when a male rat bonds with his mate. He is more caring toward her but also more likely to attack a stranger who threatens her.[5] This relationship between social bonding, oxytocin, and aggression is seen across mammals. This means the same moment a mother polar bear is the most loving—when she is with her cubs—is the same moment when she is the most dangerous. Threatening her cubs, even unintentionally, transforms her into the stuff of nightmares. Her love for them makes her willing to die protecting them.

As our species was shaped by self-domestication, our increased friendliness also brought a new form of aggression. A higher availability of serotonin during human brain growth increased the impact of oxytocin on our behavior.[6] Group members had the ability to connect with one another, and the bonds among them were so strong, they felt like family. With minor changes to the connections in the brain's theory of mind network early in development, caregiving behaviors extended to a variety of social partners beyond their immediate family.[7] With this new concern for others came a willingness to violently defend unrelated group members or even intragroup strangers. Humans became more violent when those we evolved to love more intensely were threatened.

UNIVERSAL DEHUMANIZATION

One of the fundamental principles of social psychology is that people prefer their own group members.[8] We can become highly xenophobic when responding to a stranger from a rival group—particularly during times of conflict—and it takes very little to trigger this group psychology.[9] Dividing strangers into groups based on almost any arbitrary difference can lead to antagonism: giving one group a yellow armband and leaving the other group without; separating people according to whether they have blue or brown eyes; calling one group the "overestimators" and the other group the "underestimators" of dots on a screen.[10]

No one needs to teach us to prefer those more like

ourselves. Our preference for those most like us first appears in babies.[11, 12] At nine months old, babies prefer puppets that help a puppet who is similar to them in some way—a puppet that likes the same food, for example. They also prefer a puppet that harms puppets that like different food.[13] Children are more willing to enforce norms in their midst when the child violating the norms is an outsider rather than a group member.[14] By six years old, children are willing to pay a cost (in candy) to punish cheaters who are outsiders but are less likely to punish their group members.[15]

In the famous Robber's Cave experiments in 1954, Muzafer Sherif randomly divided a group of eleven-year-old white boys into two teams at a summer camp in Oklahoma. Camp counselors on each team described the other team as a threat. Within a week, the teams were burning each other's flags, raiding each other's cabins, and making weapons. This early-emerging proclivity to attribute negative characteristics to outside groups has been credited with motivating everything from discrimination to genocide.

Social scientists have traditionally called this tendency prejudice, generally defined as a negative feeling toward a group of people.[16] The human self-domestication hypothesis suggests our worst behaviors toward other groups cannot be explained as mere "negative feeling" toward others. It suggests we also evolved the ability to dampen the activity of the mental network that produces the unique features of our theory of mind. This allows us to blind ourselves to the humanity of people outside our group when we feel threatened. This blindness is a far darker force than even prejudice. Unable to summon empathy for outsiders, we feel no connection to their suffering. Aggression is permissible. Rules, norms, and morals for their humane treatment no longer apply.[17]

If our hypothesis is correct, we should find evidence that the brain's theory of mind network becomes less active when we perceive our group is threatened. This dampening should be associated with a willingness to inflict suffering upon threatening outsiders. People might vary in their predisposition to dehumanize, and levels of dehumanization might be heavily impacted by socialization, but our hypothesis predicts that *all* human brains are capable of dehumanization.[18]

THE DEHUMANIZING BRAIN

Reduced activity in our brain's theory of mind network has been linked to negative treatment of outsiders. Recall from chapter 5 that the amygdala responds to threats and its reactivity influences our brain's theory of mind network (mPFC, TPJ, STS, PC).[20] Oxytocin plays an important role in modulating this relationship. By binding to neurons in the mPFC, oxytocin amplifies the amygdala threat signal and blunts the response of the mPFC during social interactions.[21–23]

The neuroscientists Lasana Harris and Susan Fiske tested how people classify one another based on their relative warmth and competence. A warm person has good intentions while a competent person can carry them out. The two traits can vary independently. Someone can be high in competence and low in warmth, and vice versa. For example, most people see the elderly as high warmth but low competence, and rich people as high competence but low warmth.

When Harris and Fiske showed people in an fMRI

machine photographs of people in the low competence, low warmth category, like homeless people and drug addicts, they saw that the images were processed differently in the brain than those in other categories. People in the low competence, low warmth category elicited more activity in the amygdala than the others had, suggesting that people felt threatened by them and were less likely to attribute a fully human mind to them.[24]

Oxytocin's role in modulating antipathy toward outsiders has been demonstrated by administering it as a vapor that people can inhale.[25] Men given oxytocin this way during a competitive economic game were three times more likely to donate money to their group members, and more likely to want to aggressively punish outsiders for not donating enough.[26, 27]

In the most dramatic demonstration of oxytocin's effect on our behavior toward outsiders, a group of Dutch men were faced with a moral dilemma. In the experimental scenario, they were part of a team of six people exploring a cave by the sea. One of the men became stuck in the small entrance hole. If the man was not removed, the rising tide would flood the cave and everyone would drown, except for the man in the entrance hole, whose head was out of the cave above the water line. One of the people trapped in the cave had a stick of dynamite. The Dutch men were asked if they would use the dynamite to widen the entrance and save the group, even though doing so would kill the man trapped in the hole.

In one version of the dilemma, the man stuck in the hole was given a Dutch name, like Helmut, and in the other version, the man had an Arab name, like Ahmed. When the Dutch men were given nasally administered oxytocin, they were 25 percent less likely to sacrifice the trapped man if he had a Dutch name than if he had an Arab name.[28]

Reduced activity in every region of our theory of mind network has been linked to negative treatment of outsiders. The mPFC and TPJ are quiet when we punish others unfairly. Swiss Army officers played a cooperation game while their brains were imaged in an fMRI scanner. While playing, they saw either members of a different platoon or their own platoon cheat. When the officers saw members of a different platoon cheat, their mPFC and TPJ were relatively inactive, while they readily punished the cheaters. When they saw members of their own platoon cheat, these same areas became more active, and they did not punish them. Group identity, not rule violations, best predicted the activation of theory of mind networks, resulting in either tolerance or punishment.[19]

Other researchers found that giving oxytocin to people from one ethnic group makes them less likely to perceive facial expressions of fear or pain in the people from a different ethnic group.[29, 30] This decrease in the sensitivity toward the fear and pain of outsiders was consistent across all the tested ethnic groups. Finally, growing up surrounded by conflict may further exaggerate this effect. Adolescents who grow up in areas plagued by chronic ethnic conflict have higher circulating oxytocin levels and reduced empathy toward the rival ethnicity.[31]

UNIVERSAL ACROSS TIME AND CULTURE

All the living members of the great ape family, except bonobos, are typically fearful or aggressive toward others simply because they're strangers. All except bonobos have been seen to kill strangers. Our last common ancestor with the other ape species was likely highly fearful or aggressive toward strangers. All human species that subsequently evolved probably shared this trait.

While our species converged with bonobos in our friendliness toward strangers, in humans, this friendliness extends only toward some strangers. We judge strangers based on group identity. Love for our own groups enhances our fear and aggression toward strangers with a different identity.

This is consistent with what we understand about hunter-gatherers. In every hunter-gatherer group researchers have studied, men preemptively carry out raids against outsiders to protect their own groups; lethal raids are the top cause of mortality among adults.[32, 33]

While the label "genocide" did not exist until after World

Democide and Genocide 1818–2018

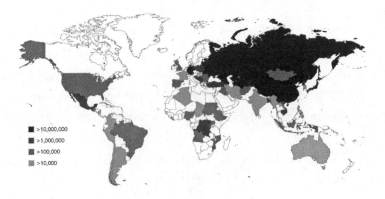

Deaths by country from democide (when governments massacre unarmed people) and genocide. From Rudolph Rommel, "Power, Genocide, and Mass Murder," *Journal of Peace Research*, 31(1) 1994.

War II, there are accounts of ancient massacres at Carthage
and Melos and of violence at the level of genocide in antiquity
in Persia, Assyria, Israel, Egypt, and the Far East.[34] Even the
modern industrialized world is susceptible to this kind of violence. Over the last two hundred years, there have been multiple major genocides on every continent except Antarctica.

Evidence for dehumanization turns up in every culture
we study.[35] The social psychologist Nour Kteily and his collaborators have recently begun to carry out a series of pioneering studies using "The March of Progress," a drawing
meant to illustrate 25 million years of human evolution.

Commissioned by Time-Life Books in 1965, "The March

of Progress" is a misrepresentation of our species' evolution
that, like "survival of the fittest," resonated with the public.
This drawing implies that evolution is a linear progression
and that humans are its pinnacle—though neither idea is
true. Even though the text accompanying the illustration
clearly stated that this was not the case, editor F. Clark Howell acknowledged ruefully that "The graphic overwhelmed
the text. It was so powerful and emotional."[36]

Despite the damage the graphic has done to the public's
understanding of evolution, Kteily realized that it could be a
powerful measure of dehumanization. He renamed it the

"Ascent of Man Scale," and surveyed more than five hundred Americans by asking what to many seems like a shocking question. First, he asked 172 Americans (mostly white) to evaluate the following statement: "People can vary in how human-like they seem. Some people seem highly evolved whereas others seem no different from lower animals. Using the image below, indicate, using the sliders, how evolved you consider the average member of each group to be," where fully human was scored at 100.

He found that, as a group, the sample he tested reported half the ethnic groups as less than fully human. Muslims were

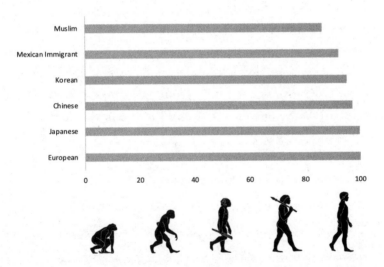

the most dehumanized, scoring a full 10 points below fully human. *Any* reported difference here goes against decades of biological research as well as modern norms of equality. Obviously all these different populations are fully human, but a significant portion of participants saw variation where none exists.[37] This dehumanization was not abstract. People who dehumanized Muslims were the most likely to sanction both their torture and drone strikes in the Middle East.

As people felt more threatened by a particular group, Kteily found that dehumanization increased. He measured dehumanization of Muslims both before and after two extreme Islamists detonated bombs at the finish line of the Boston Marathon, and he found that rates spiked by almost 50 percent after the attack.[38] The same spike occurred in the UK after a Muslim man killed a British soldier. Again, people who dehumanized Muslims the most were more likely to support drone strikes and counterterrorism efforts. Both attacks also predicted a tendency to generalize a single attacker's actions to the whole of Islam.[37]

It is not just Americans who respond this way. Kteily also looked at Hungarians' dehumanization of the Roma (formerly known as Gypsies), an ethnic group in Europe that has been persecuted for centuries. Enslaved in medieval Europe, the

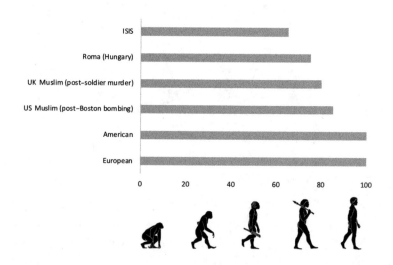

Roma were later forcibly settled and stripped of their cultural identity. The Nazis killed a fifth of their population. The majority of this population live well below the poverty line and are the target of harassment and discrimination. When Kteily and his colleagues gave the slider test to Hungarians, they dehumanized the Roma more than Americans dehumanized Muslims even *after* the terrorist attacks. In fact, Hungarians rated the Roma as somewhere between a *Homo erectus* and an australopithecine.

Similar results were found when comparing the levels of dehumanization between Palestinians and Israelis after the 2014 Gaza war. Both groups showed the same extreme levels of dehumanization of their rival group.[31, 39]

In all these studies Kteily included implicit measures of liking and disliking to test whether feelings of prejudice explained people's violent attitudes toward others better than dehumanization did. This included more implicit measures that did not require blatantly dehumanizing other groups. Kteily repeatedly found that his measure of dehumanization was best at accounting for people's willingness to inflict harm and suffering on other groups of people.[37, 40]

Kteily also found that part of our brain's theory of mind network becomes selectively activated as we judge if other people are human. The precuneus (PC) became more or less active as American participants judged whether or not individuals from different groups (Americans, Europeans, Muslims, the homeless) should be considered fully human.[41] Remember that the explosive growth of the PC was mainly responsible for the unique globular shape of our head, which appeared only *after* we separated from Neanderthals.

Common sense would suggest that it was a threat to resources, prestige, or some other economically valuable commodity that was most likely to encourage the dehumanization of another group. Maybe it was competing political ideologies of different groups or the relative status of one group within a larger society that made them more or less likely to dehumanize another. Kteily found, however, that while these elements could play a role, the element most likely to predict one group's dehumanization of another was the perception that *they* were the ones being dehumanized. This is called reciprocal dehumanization.[42, 43]

For example, Americans doubled the level of dehumanization of Arabs on the Ascent of Man scale after they were presented with a fictitious *Globe* article headlined IN LARGE PARTS OF MUSLIM WORLD AMERICANS PERCEIVED AS ANIMALS. The article also suggested that this was a view held by the majority of Muslims. Reciprocal dehumanization also informs attitudes toward peace between conflict groups. Both Israelis and Palestinians were more likely to support punitive and antisocial policies toward the other group based on the degree they felt the other group was dehumanizing their own.[42] Every population and culture that has been examined shows the same pattern of group threat leading to dehumanization of the threatening group.

WE ARE ALL SUSCEPTIBLE

I was fourteen years old when I saw Nayirah testify that Iraqi soldiers had stormed a hospital in Kuwait and thrown premature babies out of their incubators. Nayirah was only a year older than I was. Her voice breaking, she described how she saw the babies left to die on the cold floor. I knew nothing about Kuwait, but hearing Nayirah, I was horrified. I remember thinking, "Those Iraqi soldiers are animals. We have got to do something."

I was not the only one. President George H. W. Bush, as he sold the need for intervention, cited the incubator story ten times in the weeks that followed. Seven senators cited the story when casting their votes to go to war, and the motion passed by only five votes. Comparisons of Saddam Hussein to Hitler abounded. Many credit Nayirah's testimony for galvanizing the American public against Saddam and his invading forces.

In the end it turned out that Nayirah's testimony was fabricated. She was the Kuwaiti ambassador's daughter, and her testimony was part of a campaign run by the PR firm Hill and Knowlton to sway the American public's support to defend Kuwait. Hill and Knowlton knew exactly which buttons to push to create support for the largest military action since Vietnam.[44, 45]

Most of us would respond to a child in distress, help comfort a colleague whose spouse died, or take care of a sick relative. We have all made friends with people who were once strangers. We have tremendous potential for compassion and we evolved uniquely to show friendliness to intragroup strangers. But our cruelty to one another is connected to this kindness. The same parts of our brain that tamed our nature and facilitated cooperative communication sowed the seed for the worst in us.

The Uncanny Valley

In 2007, a group of Baka pygmies, one of the last remaining hunter-gatherer groups living in the rainforests of the Congo Basin, were housed in the Brazzaville zoo.

The Bantu are the dominant ethnic group in Congo, and their treatment of pygmies has often been inhumane. The Bantu word for pygmies is *ebaya'a,* which means "strange inferior beings,"[1] and the Bantu sometimes have kept pygmies as slaves. During the Congo War of 1998, Bantu soldiers even hunted and ate pygmies as if they were animals.

The Baka pygmies were musicians, and the Congolese government had brought them to play in a music festival in Brazzaville. While the other musicians were put up in hotels, twenty pygmies, including women with babies, were crammed into a tent in the zoo. The government insisted the pygmies would be more comfortable there, since the zoo was closer to their "natural environment."[2]

It was not the first time pygmies have been displayed in a

zoo. In 1906, a pygmy named Ota Benga was put on display in the Monkey House at New York's Bronx Zoo. Displaying indigenous people as exhibits was popular in Europe and America in the nineteenth and early twentieth centuries. Well-dressed visitors would marvel at "backward races"[3] in human zoos, sometimes displayed alone, sometimes in groups of hundreds, sometimes in replicas of their home environments and sometimes in cages with animals.

Ota Benga was twenty-three years old and stood four feet eleven inches tall. He weighed 103 pounds. His teeth were filed to points, and while on display, he wore only a loincloth. On Mondays, the loincloth was washed, leaving him naked.

The enclosure gave Benga nowhere to hide from the thousands of visitors who flocked to the zoo to see him. He had to ask a keeper, who did not speak Benga's language, to let him in and out of his enclosure, and the keeper might or might not comply. On the rare occasions when Benga was allowed to wander around the zoo, he was harassed and hounded back to his exhibit. Zoo visitors loved to watch Benga play with a young chimpanzee and marveled at how similar Benga and the chimpanzee were. It seemed to them that Benga and the chimpanzee spoke the same language and that sometimes it was hard to tell one from the other.

When Benga was released from the zoo, he was sent first to live at an orphanage and then to work in a tobacco factory. He had his teeth capped, bought American clothes, and learned English. He left no written record of his experiences. But we do know that at thirty-three years old, Ota Benga built a ceremonial fire, chipped the caps off his teeth, and shot himself in the head.

Since medieval times, monkeys from the Americas, Asia, and Africa have been a status purchase for the rich[4] and have been admired for their love of mischief and trickery. Philosophers as far back as Aristotle had remarked that monkeys could be the missing link between humans and beasts, but no one seemed threatened by this close association.

The great apes were different. Until a few hundred years ago, they were not much more than legend to people who lived far from their natural habitat. In the seventeenth century, explorers brought back tales of gargantuan apes who walked upright and used weapons. All great apes were called *orang hutans*, Malaysian for "men of the forest," and were often confused with pygmy people and with fantastical monsters who lived in the uncharted jungles of the "dark" African continent.[5]

It was not until the eighteenth century that a number of

great apes, both dead and alive, began to arrive in Europe.
They were dissected by anatomists and gawked at by royalty.[6]
These were not pint-sized monkeys who could be dressed up,
ridiculed, and kept in captivity with a collar and a leash. They
were hulking black apes who, when grown and standing on
two feet, could look a man in the eye and hurl him across the
room.

The roboticist Masahiro Mori suggested that the more
human robots looked, the more appealing they would be to
us. But he also said that there was a degree of similarity, a
point at which robots were almost, but not quite, human—at
which point they would elicit feelings of eeriness and revul-
sion. Mori called this place the uncanny valley.[7]

"The uncanny valley" almost certainly describes what

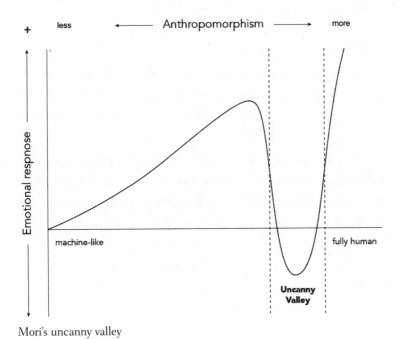

Mori's uncanny valley

Europeans felt seeing great apes for the first time. They wrote about apes in fascination and horror—as grotesque, degenerate mirrors of humans, with outrageous sexual drives and a taste for destruction. Some people speculated that great apes were the result of an unnatural union between people and monkeys.

When the great Swiss taxonomist Carolus Linnaeus attempted, in the eighteenth century, to put great apes in the same class as humans, other scientists protested and were compelled "to defend the rights of mankind and to contest the ridiculous association with the true ape."[8] This debate was reignited continually through the nineteenth century, most notably after the publication of Darwin's *Descent of Man*.

Simianization, from the Latin *simia*, "ape"—the comparing of people to apes or monkeys—is a common form of dehumanization. Great apes are a perfect tool for this tactic, because unlike other animals invoked to demean people, such as rats, pigs, and dogs, great apes fall into the uncanny valley, invoking in people deep feelings of discomfort and even disgust.

As early as the fourteenth century, Europeans described Ethiopians as having monkey-like faces,[9] but it was during the slave trade of the fifteenth to nineteenth centuries that the comparison between black people and great apes gained traction. Millions of people had been shipped across the Atlantic from Africa by the time great apes were shipped to Europe for the first time. For most of the seventeenth century, many European elites claimed they could not tell the difference between a human pygmy, a bonobo, and a gorilla.[10]

European scientists didn't know where to put great apes on their mistaken evolutionary ladder. They had put white people at the top, and the obvious resemblance of great apes to humans meant that the logical step, as Linnaeus and Darwin suggested, was to group all humans and great apes in the genus *Homo*.[8]

The strict social hierarchy of the day made this proposition too difficult for many to accept. To make our relationship to apes more palatable, nineteenth-century anthropologists inserted another rung on the ladder. The anthropologist James Hunt wrote in 1864, "the analogies are far more numerous between the ape and the Negro than between the ape and the European."[11] If apes were the intermediary between humans and the animal kingdom, black people could be the intermediary between white people and apes.

This view had the advantage of solving another dilemma, which was how to reconcile the horrors of the slave trade with the morality of upper-class elites. Simianization allowed them to morally justify excluding black people from the life, liberty, and happiness they insisted were the inherent rights of all humans.[12]

Simianization did not end with the slave trade, nor was it directed only at Africans. The Irish were simianized in both Britain and America in the nineteenth century, while the Japanese were simianized during World War II. Germans, Chinese, Prussians, and Jews were all simianized at some point in the buildup to the twentieth century's major conflicts.[13]

But while these characterizations went out of style, black people in America continued to be portrayed as apes—often with a lust for women or blood. One of the most popular manifestations of the sex-crazed ape gone mad was the 1933 movie *King Kong*. In retrospect, the racial undertones are obvious. A white woman goes to a jungle island, where she is met by black savages under the thrall of an enormous black gorilla. The gorilla takes an unnatural sexual interest in the white woman, and the white woman brings the black gorilla back to white civilization, which he is unable to appreciate. White men kill the black gorilla before he destroys white civilization, the white woman falls helplessly into the arms of the leading white man, and the natural order is restored.[14]

In 1933, nine black youths were arrested for raping two white women on a train in Alabama. The accusation was false, but on barely any evidence, eight of the nine boys were sentenced to death by electric chair. The youngest, who was twelve, had a hung jury between the death sentence and life imprisonment. In a linocut from the period, one of the boys

clutches the limp body of a naked white woman in a clear reference to *King Kong*.

Even after World War II, during the civil rights movement, cartoons of apelike black men making advances on white women were common. NO NEGRO OR APE ALLOWED INSIDE BUILDING read one sign outside a country store in Calhoun, South Carolina, in 1959.[15]

Culture is frequently invoked to explain this tendency to simianize other people.[16] Cultures are infinitely plastic and as a result are vulnerable to erroneous beliefs, abusive norms, and dubious morals. Often attributed to ignorance or economics, cultural decay should be reparable.

After World War II, scholars concluded that certain cultures were more likely to commit genocide. Germany's hierarchical culture was said to make its citizens more susceptible to authority.[17] Some said that "important realms of German society were fundamentally different,"[18] some that "mendacity has become an integral part of the German national character."[19] Japanese war crimes during World War II were attributed to a "morally bankrupt political and military strategy, military expediency and custom, and national culture,"[20] or to the Japanese "belief that the strength of the Japanese soul would compensate for material weakness."[21] The rape of millions of women in Berlin occurred because "traditional Russian culture harbors deep strains of authoritarianism"[22] as well as patriarchy and habitual binge drinking.[23]

Many social scientists have attributed the postwar decline in prejudice, especially blatant racial prejudice, to the victory of progressive Western culture. According to this narrative, when America became a superpower, the moral compass of the world swung true north. There was now an "absence of intentional discrimination"[24] and a "decline in overt and covert racism" in the United States.[24–27]

Some argue that racism was relatively new "and did not infect Europe itself prior to the period between the late medieval and early modern periods."[26] In America, the civil rights movement "smashed the legal apparatus of segregation and political exclusion." Others argued that the increase in tolerance was the result of increased knowledge. "Racial prejudices now lack the intellectual and cultural supports they had in the past," argued one researcher. "Claims of white racial superiority have been thoroughly debunked by geneticists and biologists and have been politically and socially marginalized as a result of their strong association with Nazi fascism and the Holocaust."[28]

In 2000, some social scientists declared that the culture of racism was dead in America, at least the kind of racism that led to lynchings, segregation, and internment camps. "Old racism," or negative feelings toward black people and the belief that black people are inferior to white people "has declined substantially over time." Political scientists argued that in the post-Obama era, racism was no longer a factor in political decisions, [29] and "Whites refusing to vote for black candidates has finally gone the way of segregated water fountains."[30]

THE CULTURE OF NEW PREJUDICE

However, in what the psychologist Phillip Goff calls the "Attitude, Inequality Mismatch," minorities who live in a supposedly postracial society still suffer huge inequality in employment, education, housing, income, and health. "Black and Asian Britons . . . are less likely to be employed and are more likely to work in worse jobs, live in worse houses and suffer worse health than white Britons."[28] "Of the various national groups in Germany, Turks remain where they have always been, in the forefront of hostility."[31] "Compared to other social groups in Australia, Aborigines experience disproportionately high levels of unemployment, poverty, imprisonment, and illness."[32]

This disparity is especially apparent in America's prison system. During the "War on Drugs" in the 1990s, more people went to prison for longer terms. The United States now has a higher incarceration rate than any other country, including China, Russia, or Iran, and although black people make up only 13 percent of the total U.S. population,[33] they account for 40 percent of the prison population.[34] In cities like Washington, D.C., on any given day in 1998, 42 percent of all black men living in the city were under correctional supervision. In Baltimore it was 56 percent.[35]

To explain this mismatch, researchers proposed that old prejudice (the kind that can lead to genocide) has been replaced by new kinds of prejudice. There is "near consensus among scholars that more modern forms of prejudice have generally displaced 'old-fashioned forms of racial bias.'"[36] Now, racism is "subtle,"[37] "diffuse,"[38] and "path dependent,"[24] and it can go by the name of "symbolic," "aversive," "modern," or "covert"[39] racism.[40]

Others have suggested that the current problems facing black people are largely attributable to their own moral failings. "It is simply a fact that blacks, and particularly young black men, engage in lawless conduct, very much including violent conduct, at rates (by percentage of population) significantly higher than do other racial or ethnic groups," writes Andrew McCarthy in the *National Review.*[41, 42]

At the end of World War II, when the full scale of the Nazis' "final solution" was revealed, what stunned people most was its bureaucratic efficiency. But large-scale atrocities were not seen in Nazi Germany alone. There were the Japanese rape of Nanking, the Hungarian death marches, the Russian rape of Berlin, the Romanian pogroms. Psychologists could not account for it. It was convenient to assign most of the blame to a few psychopathic leaders, but the sheer scale of the carnage made it unlikely that it was the work of just a few bad apples. The field of social psychology emerged primarily to understand what made ordinary people do terrible things.

Three dominant explanations came out of their work: prejudice, a desire to conform, and obedience to authority. Gordon Allport described prejudice as "an antipathy based on faulty and inflexible generalization."[43] It starts young, he maintained, and is stubbornly persistent. Children are exposed to the prejudice of their parents and family members, and as children's sense of identity strengthens, they start to develop an attraction to their group and repulsion to other groups. According to Allport, and a generation of researchers who followed, prejudice was the root cause of social, political, and economic inequality. To reduce prejudice, remedial measures should focus on cultural influences that shape prejudice and might lead to its reduction.

To Allport's theory of prejudice, Solomon Asch added the human desire to conform. Asch was a Polish Jew who survived the brutal invasions of the Russian and German armies during World War I, and, at thirteen, immigrated to the United States in 1920 with his family.

Asch wanted to know why millions of people placidly accepted the Nazi regime and went to their death, or stood by while friends and neighbors were murdered in front of them. Asch became immersed in "the ways in which group actions become forces in the psychological field of persons."[44]

His most famous conformity experiment was simple. Asch showed ten people sitting in a room together two cards:

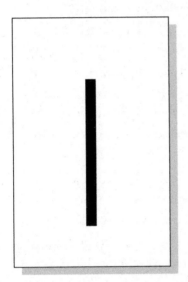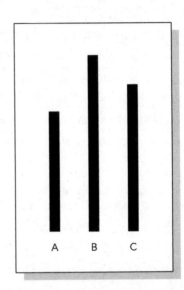

Then he asked them whether the lines on the right card were shorter or longer than the line on the left card. Nine of the ten people worked for Asch, and all of his employees gave the same wrong answer. The question was what the tenth person, hearing their responses and unaware of the purpose of the experiment, would do. If they chose the correct answer, they had to disagree with the majority of people in the room.

Asch found that 75 percent of the time, people sided with the incorrect majority opinion.[44]

Almost a decade later, Stanley Milgram, a student of both Allport and Asch, became fascinated with the trial of Adolf Eichmann, the Nazi who had organized the transport of millions of Jews to their death in concentration camps. Milgram noted that a journalist who had attended Eichmann's trial described him as an "uninspired bureaucrat who simply sat at his desk and did his job."[45] This led Milgram to perform his famous experiments testing the limits of our desire to be obedient to authority.

The picture seemed complete. While cultures of racism, moral systems, education, and economics all have a critical role in shaping group behavior, it was prejudice, conformity, and obedience to authority that became the dominant psychological explanation for the horrors of World War II. But the worst human weakness was missing from the analysis.

A year after Milgram published his famous monograph on obedience to authority, the developmental psychologist Albert Bandura published his pioneering experiment on dehumanization. Bandura wanted to find out whether ordinary people would be cruel, not in deference to someone else, but because they shared responsibility for the decision to punish. Bandura thought people would be more cruel when the decision was distributed across several people, so that the cruelty could not be traced to any one individual.[46]

People in the experiment were given the role of supervisors and were asked to manage the training of workers by using electric shocks. The main job of the supervisor was to decide the level of shock—from mild to severe on a 10-point scale—each time the workers gave a wrong answer.

Some supervisors were told they were solely responsible for determining the shock level, while others were told their decision would be averaged across several supervisors. The stated goal was always the same—to improve the performance of workers by increasing the number of correct answers. As Bandura predicted, supervisors who didn't believe they'd be held solely accountable for their actions tended to choose more intense shocks.

But there was one crucial manipulation. Just before the training began, the experimenters left the room and "forgot" to turn off the intercom between their room and the supervisors' room, so the supervisors could overhear the experimenters talking about the workers. Some supervisors heard the experimenters describing the workers as "perceptive" and "understanding" while others heard the experimenters describing the workers as "rotten" and "animalistic."

To Bandura's surprise, the subtle dehumanization of the workers had a much stronger effect than the dispersal of responsibility. While supervisors used only the mildest shocks on the workers who were humanized, supervisors gave shocks of double or even triple the voltage to workers who were dehumanized.

Even more alarming, when shocking the workers did not improve their performance, the supervisors of the humanized workers began decreasing the shock intensity, while the supervisors of the dehumanized workers *increased* the shock intensity. When Bandura asked the supervisors whether the workers' punishment was warranted, more than 80 percent approved of punishment of dehumanized workers, while only 20 percent approved of punishment when the workers had been humanized.

Not only were people able to absolve themselves for hurting someone who'd been dehumanized, they believed these subjects were less sensitive to pain and swayed only by the application of more shocks. Bandura concluded that dehumanization is central to explaining human cruelty.

Every psychology student knows about Milgram's obedience to authority experiments, but few have heard of Bandura's dehumanization experiments. Even among researchers, Milgram's work has been cited almost twenty times more often than Bandura's. We tend to consider the overt dehumanization of other groups a relic of a distant past—far beyond the pale of our civilized modern societies.[47] Instead, the focus has shifted toward interventions against the more covert forms of "new prejudice."

But understanding why we dehumanize others is crucial for understanding human cruelty.[48] Nowhere is this more clear than in Phillip Goff's research on the treatment of black people by the U.S. justice system.[49, 50] Black children receive adult sentencing at 18 times the rate of white children in the United States. They account for 58 percent of children receiving adult sentencing.[51]

Goff also found that when a black defendant was described in the media with words that people associated with apes, like "hairy," "jungle," and "savage," these black defendants were more likely to be executed by the state. Prejudice, Goff argues, is not enough to account for the disproportionate number of black children being given unusually cruel punishments. It can't account for the link between dehumanizing language and sentencing. Nor can prejudice predict extreme violence, such as genocide, toward black people.[50]

Goff points instead to dehumanization, specifically simianization. Simianization of a person or a group of people can lead to moral exclusion and the denial of basic human protections. It explains existing racial discrepancies in America better than prejudice does.

There is no shortage of simianization today. It happens to even the most famous and powerful African Americans.

 **Linda Kimball**
19 mins ·

You bunch of piece of shit porch monkeys need to stand up. Did I just call you porch monkeys? I sure did! Your ignorant sorry ass entitled pieces of shit and yep your ALL black! The civil war was fought and won a few hundred years ago cut the crap!!
Oh and calling you a monkey was actually in insult to apes since they are intelligent creatures!

Black athletes are often described in apelike terms, like "aggressive," "massive," "monster," "huge," or "explosive," while white players are more likely to be described with words like "intelligent," "commitment," or "overachiever."[52] In 2006, a spectator called Dikembe Mutombo a "monkey" at an NBA game. In 2014, a spectator threw a banana at the Brazilian soccer player Dani Alves. In 2008, LeBron James was photographed by Annie Leibovitz for the cover of *Vogue*. He was the first black man ever to grace the cover. Unfortunately, the image, which depicted a screaming James clutching the slender waist of the white supermodel Gisele Bündchen, strongly resembled the 1933 poster for *King Kong*. In 2017, the players from the National Football League who peacefully protested American racial injustice by kneeling during the national anthem were frequently simianized by their critics.

Even presidents are not immune. During Barack Obama's campaign, monkey T-shirts and monkey dolls were produced. A bar owner in Georgia sold T-shirts depicting Curious George eating a banana with OBAMA '08 written underneath.[53] This persisted throughout Obama's presidency. In 2009, a cartoon in the *New York Post* showed a dead chimpanzee with three bullet holes in front of two policemen. The speech bubble above the policemen says, "They'll have to find someone else to write the next stimulus bill."[54] These comparisons were extended to the rest of Obama's family.[55]

Some might immediately assume that this kind of simianization of black Americans occurs among older, white, rural, Republican men without a college degree.[56] But the story is not so simple.

The political scientist Ashley Jardina surveyed two thousand people, whom she carefully sampled to obtain a representative group of white Americans across every demographic.[57] She showed them Kteily's Ascent of Man scale and asked them how evolved black people were compared to white people.

On average, white people responded that black people were less evolved than white people, or closer to apes, on the evolutionary scale. For example, of the respondents, 63 percent rated white people as fully evolved, while only 53 percent of respondents rated black people as fully evolved.

When Jardina broke down the responses by demographic groups—Democrats and Republicans, conservatives and liberals, men and women, high income and low income, Southern and non-Southern, young and old—a portion of *all* groups of whites rated black people as less evolved, and more apelike, than white people. The degree to which each of these groups dehumanized black people varied and was almost always small, compared to the degree that Kteily observed Muslims and Roma dehumanized, for example, but the effect was present in every white demographic.

To confirm her results, Jardina asked the same survey respondents whether they strongly agreed or disagreed with the idea that blacks can be savage, barbaric, or have as little self-restraint as animals. Only 44 percent of respondents strongly disagreed with these characterizations of their fellow Americans. The majority of white people ranged from disagreeing somewhat to agreeing strongly. Jardina also gave people the opportunity to comment, and these were some of their responses:

"I consider blacks to be closer to the animal kingdom. They are faster, stronger and more athletic than any other race. They also lack the intelligence and morals that other races do [sic]."

"The overall way that people carry and conduct themselves. Some acting on a spectrum from almost animalistic to others acting quite civilized."

"WHAT RACE HAS THE HIGHEST RATE OF MURDERS, MURDERS THAT HAVE NO REMORSE. PEOPLE WHO ACT LIKE ANIMALS [sic]."

"When I started looking into it," said Ashley Jardina in 2017, "it did not look like new and subtle racism. It looked like black people were being denied their humanity."[57]

Nor can this kind of blatant simianization be explained by a lack of education. In 2016, the psychologist Kelly Hoffman found that 40 percent of second-year medical students, including blacks, whites, Hispanics, and Asians, believed that black people's skin is thicker than white people's.[58]

This misconception supports a myth that has persisted since slavery—that black people are less susceptible to pain. Medical students who think black people have thicker skin are less likely to treat black people adequately for pain.[58] Doctors are more likely to underestimate the pain of black patients who visit an emergency room. Black people with fractured limbs are less likely to receive pain medication, as are black cancer patients, or black people with migraines or back pain.[59] Even black children with appendicitis were less likely to receive pain medication than white children.[59]

FROM THREAT TO VIOLENCE

In every society, children are afforded more protection than adults. They are judged to be more innocent, less of a threat, and more deserving of care.[50] And yet when Phillip Goff showed photos of black children to white university students, he found that they tended to overestimate the age of black children by around five years. This means that when a black child was thirteen, the students believed that that child was already eighteen years old—old enough to be tried in court as an adult.[50] The same students did not overestimate the age of white children.

In another of Goff's experiments, a photo of a black or a white child was paired with a scenario, such as "Kishawn Thompkins was arrested and charged with cruelty to animals. He attempted to drown a neighborhood cat in his backyard." Goff found that not only did people judge the black child to be older, they judged him to be more culpable for his crimes.[50]

Goff believes these tendencies are connected to the steady stream of police officers accused of using unnecessary force against black children. Given access to records from police officers from Chicago, he discovered that almost half the officers had used some kind of force against a minor. "Use of force" ranged from using wrist locks to wielding weapons. Goff found that the officers who used the most force against children were most likely to simianize black people. Standard measures of prejudice did not predict their use of force.

Jardina found that people who rated black people as more apelike than white people were more likely to support the death penalty.[57] When a sampling of white people were told "most of the people who are executed are African American," their support for the death penalty increased.[60] The lawyer Saby Ghoshray has said that "who lives or dies is based on the defendant's success in becoming humanized in the eyes of his peers."[61]

RECIPROCAL DEHUMANIZATION

Groups who perceive they are being dehumanized will dehumanize in turn. Just as Israelis and Palestinians were more likely to dehumanize one another if they were told the opposite group saw them as less than human, the human self-domestication hypothesis predicts that black people will also reciprocally dehumanize groups they feel threaten their own.

Experimental evidence does suggest that blacks and whites show more empathy for physical pain in strangers of the same race. In one study,[62] black people saw a photograph of a black or a white hand being pricked with a needle in the sensitive area between the thumb and forefinger. They had a stronger empathetic response to seeing a black than a white hand pricked. In white people, it was the reverse.

In another study, a representative sample of Americans were asked to judge how evolved other groups of Americans are using the Ascent of Man scale. When they asked them to judge others based on their race and religion, white, Asian,

Latino, and black people heavily dehumanized Muslims. A proportion of white and black people also dehumanized one another.[63] This is consistent with what we would expect if reciprocal dehumanization occurs universally.

The human self-domestication hypothesis helps us explain both our friendliness and our potential for cruelty. Our capacity to dehumanize outsiders is a by-product of the friendliness we feel toward people who seem to be members of our own group. But unlike floppy ears or a multicolored coat, this by-product can have cataclysmic consequences. If we see someone unlike us as a threat, we are capable of unplugging them from our mental network. Where there would have been a connection, empathy, or even compassion, there is nothing. When our unique mechanism for kindness, cooperation, and communication shuts down, we have the potential for horrific cruelty. This tendency is only magnified and accelerated in the modern world of social media. Large groups can move from expressing prejudice to reciprocally dehumanizing one another with frightening speed.

BREEDING BETTER HUMANS

Whenever I give a talk on human self-domestication, someone always asks "Can't we just breed humans to be friendlier?" It seems obvious that if the secret to our species' success was an increase in friendliness, we should simply be able to select ourselves to be friendlier still. If you can breed a calm temperament and friendly disposition into a dog or a fox, why not a human? Following this logic, why couldn't you breed any other trait you wanted, eliminating the darkest parts of our nature, one by one?

Unfortunately, all roads down this path usually lead to eugenics. When the English scientist Sir Francis Galton coined the term "eugenics"—from the Greek *eu,* "good," and *genos,* "stock"[64]—the idea of selective breeding in humans had been around for thousands of years or more. Plato wrote that reproduction should be controlled by the state. Roman law ordered that deformed children be put to death. Hunter-gatherers throughout the world, from Inuits to the Ache, killed children who had physical or obvious mental impairment.

By the turn of the last century, eugenics was seen as the cutting edge of science, the solution to all the world's problems. It could take the form of preventing people from reproducing, for example through indefinite incarceration, or sterilization, which had gone from a complicated surgery to a quick outpatient procedure.

From 1910 to 1940, Americans heard about eugenics regularly. Teachers, doctors, political leaders, and even religious leaders used eugenics in lessons and conversations.[65] Politicians ran as "eugenics candidates," baseball stars gave speeches on the subject, schools and universities included it in their curricula, and Christian Women's Temperance Unions held "Better Babies" competitions. Victoria Woodhull Martin, the first American female presidential candidate, wrote that the "first principle of the breeder's art is to weed out the inferior animals."[66] The question was, who were these inferior animals?

One obvious target was criminals. In the early twentieth

century, criminals were thought to be inherently violent degenerates, predisposed to express the darker qualities of human nature.[67] It was an imperative of the eugenics movement to stop these aggressive criminals from reproducing, since criminality was thought to be a part of someone's nature and therefore able to be passed down through generations. Not surprisingly, the first eugenic sterilizations were performed in prisons.

Madness was also seen as inherently violent. As the eugenics movement gained popularity, its focus shifted from violent criminality to other types of mental illness. People with epilepsy, schizophrenia, dementia, or an IQ under 70 were all victims of "bad genes" and were deemed a threat to the integrity of future generations.

However, those at the forefront of the eugenics movement felt most threatened by people exhibiting a different kind of mental illness—those who could almost pass for normal but were seen to be lowering the collective intelligence of the population by passing on their mental deficiencies to the next generation. They landed on "feeblemindedness" as a catchall term for anyone who was "undesirable." It was applied to promiscuous women, poor people, black people, illegitimate children, single mothers—the list is so long that it is a wonder that any group escaped stigma.

In total, more than sixty thousand people were sterilized in the United States. You were probably alive during the last forcible sterilization, which was in 1983. Although the United States sterilized one-seventh the number of people that Nazi Germany sterilized, the U.S. sterilization program went on six times longer.

America's sterilization program was emulated around the world. Eugenics societies were founded in forty countries, and countries including Denmark, Norway, Finland, Sweden, Estonia, Iceland, and Japan passed sterilization laws.[68] Nazi officials consulted with high-ranking members of the sterilization program in California,[65] and when they returned to Germany and proposed their own sterilization laws, they referred to the United States as an example of what these kinds of laws could accomplish.

Eugenics was always doomed to fail—and not just because it is morally repugnant. Even though, in the foxes, selection against aggression seemed so easy, the foxes are a case of extreme selection. In many generations only 1 percent of the experimental foxes were allowed to breed, based on whether they approached a human.[69] In the Upper Paleolithic, when our species experienced selection for friendliness, our population was tiny—probably less than a million—and selection took effect over many tens of thousands of years.

Today, with more than seven billion humans and counting, to create the selection pressure equivalent to what the foxes experienced, more than 6.9 billion people would not be allowed to have children. Even then, there is no simple method to measure human friendliness the way there is in foxes. Worse, for the selection to work, you would need to identify people with genes known to be related to the type of friendliness you hope to promote; selecting people based on differences in friendliness driven by environmental factors would change nothing over generations.

It is impossible to selectively breed humans based on a set of genes related to even a relatively simple physical trait like height. Even though most people are between 5 and 6 feet tall, there are almost 700 genes involved in determining height in humans, and these genes account for only 20 percent of the variance in our final height (environmental and other factors account for the rest).[70]

Behavioral traits are far more complicated. For any behavior, there are thousands of genes involved, and they all interact and work together. Any one gene explains only a tiny fraction of variation in behavior.[71] We still have no idea how to determine which human gene networks are related to which types of human behavior. It impossible to identify people with the relevant genes that cause the type of friendliness we might want to select for. Intentional breeding for friendliness is clearly not the way forward.

Since our tools and projectile weapons made us the top predators of the Ice Age, we have embraced technology with barely a backward glance. Today, our ability to create an expansive network of innovators is driving another unprecedented explosion of technology. Could technology be the key to taming our dark side?

The dizzying pace of innovation is sometimes called the rate of accelerating change. For example, transistors, tiny switches triggered by electricity, power much of our technology. In 1958, the first computer chip had two transistors. In 2013, a chip had 2.1 billion[72]. In the 1980s, when the Internet went from 20,000 to 80,000 nodes over a two-year period, hardly anyone noticed. A decade later, when it went from 20 million to 80 million nodes in the same amount of time, it affected everyone.[73] In 2004, we sequenced the first human genome for hundreds of millions of dollars. Now machines can sequence more than eighteen thousand genomes a year for $1,000 each.[74] The futurist Ray Kurzweil predicts that in the next hundred years, we will experience the equivalent of twenty thousand years of progress.

With technology flooding every facet of our lives, it is natural to assume that new technologies will be running our societies even better in the near future. The Millennium Project[75] is a think tank that ranks the fifteen biggest global challenges every year. For almost every challenge, the Millennium Project proposes solutions involving technology. Climate change causing havoc? Switch to renewable energy and retrofit fossil fuel plants to reuse CO_2. Overpopulation bursting the earth at the seams? Build eco-smart cities, grow steak in a petri dish from stem cells, and genetically engineer high-yield drought-resistant crops. Need to make education universal? Develop scalable software for children anywhere in the world to teach themselves reading, writing, and math online in eighteen months flat.[76]

But as Tim Cook, the CEO of Apple, said, "technology alone isn't the solution. And sometimes it's even part of the problem." Because technology is, and always has been, a double-edged sword. The projectile weapon that we used to cooperatively hunt mammoths could also be used to kill our fellow humans. Nuclear power could be a critical solution to our energy crisis if we manage not to start a nuclear war. Self-driving cars will save a hundred thousand lives a year, until terrorists hijack the network and kill a hundred thousand people in a series of crashes. The Internet was an amazing tool for human advancement, until foreign governments used it to sway democratic elections.

In order for technology to be used as a force for good, it needs to be developed in anticipation of the best and worst of human nature, but it almost never is. Solutions for an even friendlier future will need new technologies, but they will not be enough to tame our dark side. Our social problems require social solutions.

8

The Highest Freedom

We did not evolve to be despots. We evolved to live in small bands of hunter-gatherers who valued only social currency and ostracized or killed anyone who tried to monopolize power. For thousands of generations, these egalitarian groups emigrated to all four corners of the world as every other human species disappeared.[1, 2]

The seeds of despotism were sown with the first crops.[3] When we began to produce and store food in large quantities, our societies grew. People worked together to monopolize resources, and the mechanisms that had kept despotism in check among small groups of hunter-gatherers began to fail. Autocrats who would have been exposed and punished in groups of one hundred people could now hide in a larger, more anonymous population, agitating subgroups within a society so they would fight one another. Tribes, kingdoms, empires, nation-states were all essentially built on this model: one group fighting another to monopolize power.

Ultimately, modern societies were organized by the whims of the most powerful subgroup within a society. Less powerful or minority groups were voiceless and relegated to serfdom or slavery. People struggled against this new order and fought wars to overturn it for thousands of years. Even if rebellions succeeded, the same hierarchical order was likely to be reestablished under the rule of a new despot from another clan, party, tribe, religion, or ethnicity. Agricultural-ists became stuck in a zero-sum game.

At the dawn of the Industrial Revolution, some Western European societies found a way out of this cycle by forming representative social systems called constitutional democra-cies. In 1689, the English Bill of Rights limited the power of the king and gave Parliament free elections and freedom of speech. Other countries slowly followed. Hierarchies re-mained, but checks on the powerful were being built into the system so that those out of power were never fully powerless. A norm was created for power sharing and compromise. Cit-

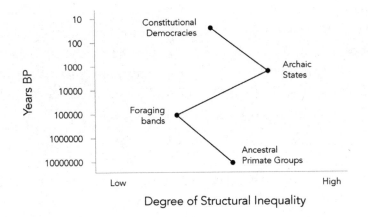

Degree of Structural Inequality

izens were not under the direction of a ruler chosen by god or pedigree, but of a citizen representing the needs of fellow citizens.[4]

Political scientists point to the steady rise in democracies since the 1970s to explain the gradual decrease in violence and the unprecedented peace of the last half century. While democratic countries do go to war, they infrequently, if ever, go to war with one another.[5] Even a low level of aggression between democracies is rare.[6]

The peace that comes with the establishment of democracies is different from the stability some dictators create. Democracies are built to defend human rights and maintain egalitarian principles so that even when a group falls from power or never had it to start with they are protected. Democratic countries are more likely to have better human rights records. They are more likely to support the freedoms of religion, the press, and speech,[7] all of which protect democracy's egalitarian spirit. Democracy can reduce income inequality,[8] and the pioneering democracies of the eighteenth century were the first to see substantial economic growth during the Industrial Revolution. Democracies tend to have better health care with lower child mortality and better maternal health. Democracies also spend more on education, have a higher teacher-student ratio, and have more of an incentive to lower school fees.[9] Democracy is crucial to the well-being of its citizens, and one of the main prerequisites of lasting peace.[5, 8, 10–12]

As they constructed their new government, the American Founders understood that people tend to form group identities along arbitrary lines, and they were painfully familiar with the cycle of dehumanization. More than a hundred years before neurobiology or cognitive psychology were recognized as sciences, James Madison neatly articulated a key feature of the human self-domestication hypothesis when he wrote

> So strong is this propensity of mankind to fall into mutual animosities, that where no substantial occasion presents itself, the most frivolous and fanciful distinctions have been sufficient to kindle their unfriendly passions and excite their most violent conflicts.[13]

At the time, these "mutual animosities" had thrown every European attempt at democracy into turmoil. The Founders had studied these failed European democracies closely. It seemed true equality was impossible. Thomas Paine wrote, "monarchy and succession have laid (not this or that kingdom only) but the world in blood and ashes."[14] The most powerful group would always trample the minority:

> A zeal for different opinions [has], in turn, divided mankind into parties, inflamed them with mutual animosity, and rendered them much more disposed to vex and oppress each other than to cooperate for their common good.[13]

In order to protect the minority from the "tyranny of the majority," the Founders eventually agreed to establish a strong central government to promote a national identity rather than allow each state to govern itself, as was the case in Europe.[15]

America is a republic—rather than a true democracy in which votes are won by the 51 percent majority rule. Our political system is intended to "protect all parties, the weaker as well as the more powerful."[16] To prevent people in the highly populated states from imposing their will on the more scattered populations of rural states, the U.S. Constitution established the Electoral College. To prevent one part of the government from becoming more powerful than the others, it established checks and balances: veto power, separation of powers, the House and the Senate.[15] The Founders were not afraid to talk about the flaws of human nature—Jay, Hamilton, and Madison explicitly referred to human nature more than fifty times in the Federalist Papers and designed a democracy meant to keep our darker side in check.[17]

The great American experiment is now being criticized from all quarters. The for-profit media, driven by market forces to entertain as much as to inform, focuses on democracy's flaws: the problematic Electoral College, squabbling politicians, corrupt financial interests, and a polarized citizenry. Political theorists point to aging political institutions. Our constitution has been called "dysfunctional, antiquated, and sorely in need of repair"[18] and the Bill of Rights "a grandfather clock in a shop window full of digital timepieces."[19] Days after the 2016 election, the Economist Intelligence Unit demoted America from a "full democracy" to a "flawed democracy." The political scientist Matthew Flinders wrote that "if the twentieth century witnessed the triumph of democracy," then the twenty-first century appears wedded to "the failure of democracy."[20]

Even the government is against the government. As we write, the cabinet includes an EPA head who is suing the EPA; an Energy Department head who advocated eliminating his own department; a head of the Department of Education who does not support public education; and a head of the Department of Labor who would like to replace human workers with robots.

In 2008, politicians in Texas proposed abolishing every federal agency not mentioned in the Constitution—including the Environmental Protection Agency, the Social Security Administration, and the Departments of Energy and of Health and Human Services—a proposal that has been repeated several times in different states. Grover Norquist, the founder of Americans for Tax Reform, said "My goal is to cut government in half in twenty-five years, to get it down to the size where we can drown it in the bathtub." [21]

Worse yet, very few Americans understand how their government is designed to work. A third of Americans cannot name a single branch of government, 29 percent cannot name the vice president,[22] and 62 percent do not know which party controls either the House or the Senate.[23]

At no point in history have Americans been so disillusioned with their republic.[24, 25] Most worrying is the disenchantment of young people. Only a third consider it essential to live in a democracy. A quarter believe that democracy is a "bad" or "very bad" way to run the country.[26] Another third would prefer to see a strong leader who doesn't have to bother

with elections. This kind of leader would, by anyone's definition, be a dictator.[27]

"Democracy is the worst form of government," Winston
Churchill conceded, "except for all those other forms."[28] Our
democracy remains far from perfect. But it is the only form of
government that has reliably demonstrated the ability to harness the better angels of our nature while muting our darker
side. As Thomas Paine wrote in 1776, "Here then is the origin and rise of government; namely a mode rendered necessary by the inability of moral virtue to govern the world."[14] So
far it has saved us from ourselves.

Not only are democracies difficult to establish and maintain, they can easily give way to dictators. "Democracies fail
when they are too democratic," warned the journalist Andrew
Sullivan in 2016.[29] A hyperdemocracy arises when a democracy facilitates so much intolerance that it begins to under-
mine itself. "Out of the highest freedom," wrote Plato in his
Republic, "comes the most widespread and savage slavery,"
creating a tyrant whose "first concern is always to be stirring
up various conflicts so that the people will need a leader."[30]

THE RISE OF THE ALT-RIGHT

The alt-right is a loosely defined group of people with far-
right ideologies who reject mainstream conservatism and
tend to score high on measures of either Social Dominance
Orientation (SDO) or Right Wing Authoritarianism (RWA).[31]

People high in SDO believe in the popular caricature of
"survival of the fittest." They believe that "some groups of

people are simply inferior to other groups" and that "an ideal society requires some groups to be on the top and others to be on the bottom."[32] In Western countries these people are attracted to white supremacy. They see it as imperative the group they identify with achieve dominance.

People high in RWA, who tend to identify as right-wing populists, think that people should look and behave in a certain way, that those who agree should be rewarded and those who disagree should be punished. They cherish conformity and the stability they believe it brings. They show tremendous kindness to their group members even as they respond with hatred toward those who do not conform to the ways of their group.

Although people high in SDO and RWA both tend to be extremely intolerant, their ideologies are distinct. While people high in RWA believe outsiders are threatening, people high in SDO believe outsiders are inferior. Those high in RWA conform to authority. Those high in SDO want their group to *be* the authority.[31]

The rise of the alt-right is not just an American phenomenon. As we write, it's taking place in liberal democracies across the world. In July 2016, thirty-nine countries in Europe had alt-right parties in their parliaments.[33, 34] As they have in the United States, these alt-right parties have been inciting violence against journalists, Muslims, and immigrants.

The media reports that economic anxiety is one of the main reasons for the rise of the alt-right, but Nour Kteily found that alt-right supporters were more optimistic about the current and future economy than nonsupporters.[35] This runs counter to the idea that poor rural communities are the most susceptible to intolerance.[36] The intolerance Kteily measured in alt-right supporters is not the result of personal trauma or ignorance.

The trait that people high in SDO and RWA overwhelmingly share is their extreme intolerance toward outsiders who seem to threaten their group identity. People with SDO are threatened by outsiders competing for dominance over their group, while people with RWA are threatened by outsiders who do not represent "the oneness and sameness that makes 'us' an 'us.'"[37] The threat is to the normative order and is compounded by diversity and freedom.

When those high in SDO and RWA feel threatened, they are likely to respond by dehumanizing members of other groups.

Within the alt-right, white supremacists (high in SDO)

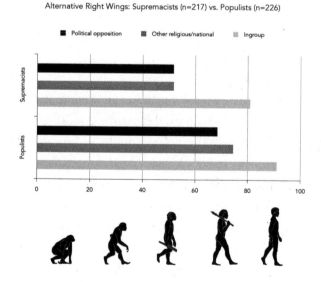

Alternative Right Wings: Supremacists (n=217) vs. Populists (n=226)

Source: Forscher & Kteily, 2017

were the most extreme dehumanizers of any group Kteily, or anyone else, has measured using the Ascent of Man schema introduced in chapter 7.

White supremacists rated feminists, journalists, and Democrats as more nonhuman ape than human. As one survey respondent wrote:

> If it were not for Europeans, there would be nothing but the third world. Racist really needs defined [*sic*]. Is it racist to not want your community flooded with 3,000 low IQ blacks from Congo? I would suggest almost everyone would not. It is not racist to want to live among your own. . . . Though media [the Jews] lie about the Holohoax, and the slave trade, Jews

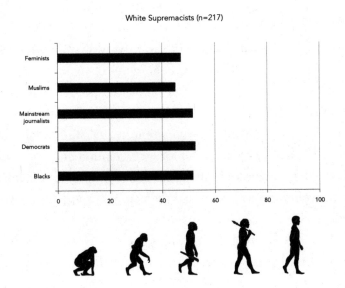

White Supremacists (n=217)

Source: Forscher & Kteily, 2017

were the slave traders, not Europeans. Many people do not even understand these simple things.

Though the historical and social misconceptions are very clear in this population, the most important finding about SDO and RWA personality is that education has very little effect.

"No one is born hating another person because of the color of his skin, or his background, or his religion," wrote Nelson Mandela. "People must learn to hate, and if they can learn to hate, they can be taught to love, for love comes more naturally to the human heart than its opposite." It is a beautiful saying, and it captures what people want to believe about intolerance—that it is a result of "closed-mindedness and ignorance"[38] and that we can teach people to think differently. "According to this wishful understanding of reality," the political scientist Karen Stenner writes, "the different can remain as different as they like and the intolerant will eventually have the intolerance educated out of them."[37]

However, trying to "educate" intolerant people might actually make things worse. Recall that when Ashley Jardina informed white people participating in her study that black people were unfairly targeted for incarceration and execution, those who already dehumanized black people dehumanized them even more and increased their support for these punitive policies. Knowledge exacerbated the problem.

Value confrontation, teaching tolerance for diversity, or priming multiculturalism can backfire.[37] These tactics seem to show their greatest effects on those who are already on the tolerant end of the spectrum. For those on the other end, prescribed multicultural sensitivity training may just entrench their intolerant ideology more deeply.[39]

Those on the most extreme ends of both SDO and RWA "will never live comfortably in a modern liberal democracy,"[37] because at its heart, democracy is designed to promote the distribution, rather than the consolidation, of power, the celebration of differences rather than similarities, and equal rights for all. Differences can be difficult to celebrate if you view your own subgroup within a nation as superior, or if these differences threaten your group's ability to conform.[37]

YOU DON'T HAVE TO BE LEFT TO BE RIGHT

Dehumanization is not simply the product of one country, economy, or culture, and the alt-right accounts for only some of the challenges to democracy.

The human self-domestication hypothesis predicts that the ability to dehumanize the "other" is a human universal and can appear across the political spectrum. People on the outer extreme of *any* political ideology are the likeliest to dehumanize their political rivals.

Think of it as target with an enlarged bull's-eye on it.

Most people in a representative democracy are in the "moderate middle." They might swing one way or the other as they respond to events, but they are fairly responsive to facts. They debate the efficiencies of the market versus government spending, reconciling capitalism with more egalitarian goals, or the balance between the need for conformity in law-abiding citizens and the nonconformity that drives innovation. But usually, the middle is able to compromise—even if it is difficult.

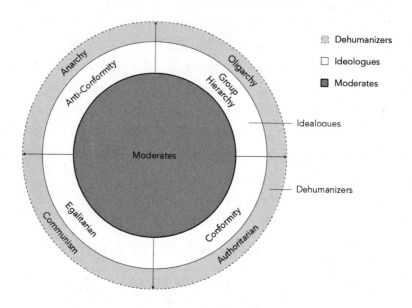

Outside the moderate middle are the ideologues. These people believe that their political views are right and all others are wrong. Ideologues are usually unresponsive to facts that contradict their political beliefs and are less interested in compromise. They create the echo chambers on social media and engage only with journalism that confirms their beliefs. If anything, they may also tend to be *more* educated.[40]

Extremists are those on the outer circle of the target. These include people high in SDO, who would welcome oligarchy (as long as their group members were the oligarchs), and people high in RWA, who would trust an autocratic leader who defended the values they feel are threatened. But these groups are only half of the picture. There are also extreme forms of egalitarianism, like communism, or the rejection of all governmental authority, like anarchism.

The human self-domestication hypothesis predicts that

all extremists on the outer ring of the target will be more likely to morally exclude—dehumanize—those who threaten their worldviews or challenge their assumptions.

But people's political beliefs are fluid. People can shift from the middle to the outer circle and back again in response to personal or political events, as they move to and from cities, grow older, or make more or less money. Politics becomes more volatile when the ideologues in the outer ring, feeling their group identity threatened, are pushed further outward toward extremism. If the threat is big enough, it can push even those in the moderate middle to the extremes.

We have already seen evidence of the universal tendency to dehumanize across the political spectrum. As discussed in chapter 7, Ashley Jardina found that in every demographic there was a subset of white people who dehumanized black people: Republicans and Democrats, old and young, women and men, rural and urban. No social or political group is immune.[41]

The extreme dehumanization practiced by white supremacists has been answered by extremists who feel compelled to respond to them with violence. The "antifa"—antifascist, or anti-white-supremacist—protesters in 2017 scrawled "Death to the Klan" on a Confederate statue, burned the Confederate flag, and carried an ax to their protest. This dynamic is not unique to any political movement, culture, or time.

The Cultural Revolution of China, Stalinism after World War II, anarchist terrorism, the French Revolution, the Japanese Empire—all forms of government can be co-opted for dehumanization and the ensuing violence. You only need to convince people they are under threat. As the Nazi leader Hermann Goering said in his prison cell at Nuremberg, "The people can always be brought to the bidding of the leaders. That is easy. All you have to do is tell them they are being attacked and denounce the pacifists for lack of patriotism and exposing the country to danger. It works the same in any country."[42]

Across time, culture, and country, the underlying psychology is always the same. To initiate the cycle of dehumanization, extremists may convince their own group that they are being dehumanized by another. As the real or perceived threat level increases, even people in the middle move away from the bull's-eye and closer to the outer circle of the target and are primed for violence against their enemies. Without a humanizing or galvanizing cause, like a moon shot or a common threat, that can be used to unite the different sides, those in the moderate middle struggle to bring extremists and ideologues back to the negotiating table.

Liberal democracy was designed to keep this darker side of our friendly nature in check. There has been much discussion about the challenges that face this form of government: debilitating debt, military overreach, fading infrastructure, misinformation campaigns, and aging institutions, to name a few. Specifically in the United States, the focus has been on a lack of civil discourse,[43] gerrymandering, arcane congressional rules that prevent bipartisanship (e.g., the Hastert rule), voter suppression, and corruption of elections through unlimited private funding.[44-47] But the self-domestication hypothesis tells us that many of these problems are just symptoms of a more fundamental challenge, the paradox of human nature: our kindness to the in-group and cruelty to those outside it.[48]

Now that we have identified the disease, we can look for the cure. Ideally, we would immunize ourselves against dehumanization, so American democracy can function the way the Founders intended. The good news is that the vaccine exists, and we know it works.

LOVE IS A CONTACT SPORT

By the time the Second World War broke out, Andrzej Pitynski had already saved several Jews by hiding them in his apartment in Poland. When the Nazis invaded, Andrzej used his job at a German firm to get passes into the ghetto and smuggle food to Jewish orphans.

When his cover was blown in 1941, he was imprisoned for two months. The guards beat him so badly they broke his jaw. After prison, Andrzej and his wife escaped to the Ukraine and rescued Jewish people working on oil refineries. The SS caught on, and Andrzej and his wife fled back to Poland. Andrzej joined the Underground Army and continued to help Jews until the end of the war.[49]

During the Holocaust, thousands of people risked their lives to help Jewish people escape persecution and death. If these rescuers were discovered, the punishment was torture, deportation, or death—sometimes for their entire families. Still, they hid Jewish people in their barns and attics, in sewers and animal cages. They looked after them for a night or for a year. They pretended that Jewish people were their nieces or nephews, or long-lost grandparents from the other side of Europe.

What made these people risk their lives while others stood by and did nothing? On the surface, nothing seemed to unite them. They were not otherwise heroic or rebellious. They were both men and women, both educated academics and illiterate peasants. They were deeply religious and complete atheists. They were rich and poor, city dwellers and rural farmers, teachers, doctors, nuns, diplomats, servants, policemen, and fishermen.[50]

When the sociologists Samuel Oliner and his wife, Pearl, analyzed the testimonials of hundreds of rescuers, they found just one common denominator: They'd all had close relationships with Jewish neighbors, friends, or co-workers before the war. Andrzej had a Jewish stepmother.[49] Stephania, who used her job to forge papers for almost two hundred Jewish girls, had a best friend who was Jewish. Ernst, who joined a resistance group when he was just fourteen, grew up with Jewish playmates.[49]

Before World War II, looking at border territories that were war zones and longstanding feuds between neighboring ethnic groups, researchers assumed that contact between different groups ignited conflict. They assumed that people felt safer in their own communities where others spoke the same language and ate the same food in the same way. Protecting cultural identity, especially to minority groups who felt disadvantaged, seemed like a priority.

Many black civil rights activists argued against desegregation. "I can see no tragedy in being too dark to be invited to a white school social affair," wrote Zora Neale Hurston in 1955.[51] They foresaw the hard road ahead for their children, as well as the firing of thousands of brilliant, caring black teachers and administrators (while white parents might tolerate having their children educated alongside black children, they certainly would not tolerate having their children educated by black teachers.) W.E.B. Du Bois wrote that "a separate Negro school, where children are treated like human beings, trained by teachers of their own race, who know what it means to be black,[52] [would be] infinitely better than making our boys and girls doormats."[53] This reasoning was used for many years to justify segregation, by both powerful majorities and disadvantaged minority groups.[52, 54]

But after World War II, researchers found that contact was the only thing that reliably reduced intergroup conflict. The best way to defuse conflict was to diminish the perceived sense of threat between groups. If groups could come together in low-anxiety situations, they discovered, these strangers would have the chance to empathize with one another. Reducing this anxiety was one of the central factors in reducing intergroup conflict. If feeling threatened turns off our theory of mind network, engaging in nonthreatening contact seems to switch it back on.

Most policies are enacted with the assumption that a change in attitude will lead to a change in behavior, but in the case of intergroup conflict, it is the change in behavior—in the form of contact—that will most likely change attitudes.

* * *

Though educating the intolerance out of people has limited effect, education does have a critical role to play through socialization. Schools and universities are ideal places for sustained friendly contact.[55] Think of Carlos and the Jigsaw program from the introduction. School desegregation in the 1960s was turbulent, and some may argue that it was not entirely successful. But in the end, interracial contact in schools did help to dissolve negative racial stereotypes. White children who went to school with black children in the 1960s were more likely, as they grew up, to support interracial marriage, have black friends, and be willing to welcome black people into their neighborhood.[56]

Even today, contact in education works. Pairs of freshman roommates at UCLA who were of different races reported more comfort in mixed-race interactions and more tolerance toward mixed-race dating. They had more mixed-race friends and were more likely to date someone of another race. White roommates paired with a black or Latino roommate were also more tolerant. The effects held steady even in seniors, years after they had lived with their freshman roommate.[55, 57, 58]

The military is another institution that is optimal for sustained friendly contact. When the U.S. Army recruited 2,500 black soldiers to fight in the Battle of the Bulge, even the more intolerant white soldiers from the South who fought alongside the new recruits came out of the conflict with more positive attitudes toward black people than white soldiers who did not.[52] This effect was also observed when the U.S. Marines desegregated in 1948.

After World War II, a U.S. housing shortage in the mid-1940s made mixed neighborhoods a necessity. White women who had friendly conversations with their black neighbors liked their black neighbors more and were more supportive of interracial housing. Not only that, but half the white tenants who lived in desegregated housing were more likely to support unrestricted access to future unsegregated housing, while only 5 percent of white people who lived in segregated housing supported this view.[59, 60]

This kind of beneficial contact can be as simple as a casual conversation, a work partnership, or a mixed classroom. It can happen organically—in a restaurant, for example—or be created artificially in a laboratory. One study found that just imagining positive contact with one of the most dehumanized groups of people—the homeless—helps people to empathize with them.[61–63] Even using humanizing words to describe people in an outside group can lead people to want to approach and make contact.[64]

So it should not be surprising that contact with imaginary characters also changes minds. Harriet Beecher Stowe's novel *Uncle Tom's Cabin* was a turning point in the abolitionist movement. A Rwandan soap opera helped reduce prejudice and conflict between Hutus and Tutsis after the genocide.[39] Though hardly a new or cutting-edge approach, storytelling is a proven method of improving our ability to empathize with people who seem like outsiders.

Best of all, contact seems to have the greatest impact on the most intolerant people. The psychologist Gordon

Hodson found that people high in SDO and RWA are the most affected by contact with a range of stereotyped groups including homosexuals, black prisoners, immigrants, the homeless, and AIDs patients. By the end of repeated interactions with members of outside groups, they start to look like the most tolerant people in the sample, as depicted below.[65, 66]

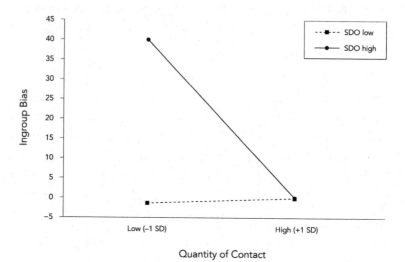

Quantity of Contact

The self-domestication hypothesis explains why we are designed for contact and how it causes a positive effect. If the members of your group are threatened, any empathy you would normally be able to establish with outsiders is blocked.[67, 68] Outsiders, feeling threatened in turn, dehumanize the first group, creating a feedback loop of reciprocal dehumanization.[69] If by establishing contact we can remove that sense of threat, even for a short time, we can create a different kind of feedback loop that might be called reciprocal humanization. Creating the conditions for contact allows for more social bonding and overall sensitivity to the thoughts of others.[70] Interaction between people of different ideology, culture, or race is a universally effective reminder that we all belong to the same group.[71]

The most powerful form of contact is true friendship, and the tolerance friendship generates seems to be contagious.[52] People are less likely to dehumanize LGBTQ individuals, for example, if they have had extended contact with people of different sexual orientation or gender identification through their network of friends.[72] Israeli and Palestinian teenagers who traveled to the United States to attend a three-week camp together were also asked to list the five people they felt closest to at camp. Around 60 percent of the teenagers listed someone from the opposite group in their top five. This high percentage predicted positive attitudes toward the opposite group as a whole.

Unfortunately, true friendship across groups can be as rare as it is powerful. According to a survey in 2000, 86 percent of white Americans know a black person, but only 1.5 percent named a good friend who was black.[73] Among black people, only 8 percent of their close friends are white.[74]

The rarity of cross-group friendships might provide a partial explanation for why so few people risked their lives to help Jewish people during the Holocaust. It certainly explains why those who did risk their lives did so without hesitation. It was not because they were particularly brave or religious or rebellious. It was because they had once, or still, loved someone who was Jewish. For these people, being human came first. Everything else was a distant second.

THE GENERAL'S GRANDDAUGHTER

On a flight to Los Angeles, I sat next to an elegant woman with short blond hair. We started talking, and she said that her name was Mary and that she worked for a nonprofit called People for People. "We bring people together from all over the world, and encourage peace through friendship," she told me.

"What got you into that line of work?"

"My grandfather, Dwight Eisenhower."

I did not expect to be sitting next to the granddaughter of the thirty-fourth American president, but Mary was easy to talk to. I told her a little about our research on self-domestication, about how friendship was our species' winning strategy but that the occasional short circuit increased our potential for dehumanization.

"Granddad never talked about the war," said Mary. "But he had a book full of photographs of the Holocaust."

Eisenhower personally visited the concentration camps. Photographs show him staring somberly at sprawling bodies and hollow-eyed prisoners.

"He said he kept the pictures because he needed to remember."

I asked Mary what it was like to have a president for your grandfather.

"I didn't think anything of it. I thought it was normal. But I do remember one thing out of the ordinary, from when I was a child."

Nikita Khrushchev, the Russian president, had come to visit the White House and given all the grandchildren toys.

He gave Mary a beautiful doll. Mary was playing with the doll on the floor when she heard someone yelling. Outside on the balcony, she saw her grandfather red in the face, shouting at Khrushchev. "I had never seen Granddad so angry." Eisenhower stormed into the room, snatched the toys from his grandchildren, and stormed out.

Those were the first, frightening days of nuclear weapons. Bombs were being built that were a thousand times more powerful than the atom bomb that was dropped on Hiroshima. People built bomb shelters in their backyards and stockpiled food for a nuclear winter.

Mary found out later that on the balcony, in full view of the children playing with their new toys, Khrushchev had pulled Eisenhower close and whispered, "I'll see your grandchildren buried."

Mary cried and begged for the doll back, and her grandfather, being a soft touch, gave it to her. But Mary always harbored an antipathy for Khrushchev, the man who had made her grandfather so furious.

Many years later, as a guest at a function for People for People, Mary looked across the room and saw Sergei Khrushchev, Nikolai Khrushchev's son, and was overcome with dread. What were the organizers thinking, inviting a Khrushchev to an event that was in honor of her grandfather?

When they were introduced, Sergei took her hand and leaned in close. "My dear," he whispered, "I hope you're not as uncomfortable as I am."

She burst out laughing, and they spent the evening telling one joke after another. From that day onward, they were great friends. Mary started working for People to People and soon became president of the organization.

"I saw how my anger and hatred could be transformed into something different," said Mary. "How just one kind word can turn an enemy into a friend. By bringing people together, we can have peace. Just like Granddad wanted."

On January 21, 2017, the day after Donald Trump's inauguration as president, more than three million people came together in the Women's March. Most of the protests were in the United States, but satellite marches were broadcast from places as far-flung as Australia and Antarctica.

A few weeks later, on February 1, 2017, 150 left-wing radicals, also known as antifa, or antifascist, protesters, arrived at the University of California, Berkeley, to protest a

scheduled lecture by the right-wing activist Milo Yiannopou-los. The antifa demonstrators were clad in black, masked, and armed with clubs and shields. They lit fires, hurled Molotov cocktails, and broke windows. Six people were injured, one person was arrested, and $100,000 worth of damage was done to the campus.[75]

Which type of protest is more likely to be successful? The Women's March made a statement, but there was no way to measure the impact of protests on their objectives. On the surface, the antifa were successful. Yiannopoulos's talk was canceled, and "antifa" became a household name. The group gained even more publicity by turning up at right-wing rallies and often clashing violently with alt-right supporters. After the white nationalist leader Richard Spencer was punched on television by a black-clad assailant, the hashtag #PunchANazi went viral and the video was set to about a hundred different songs.

Jean-Paul Sartre wrote, "the peasants must drive their bourgeoisie into the ocean."[76] The political scientist Jason Lyall found that nimble, violent rebel groups are more likely to defeat the lumbering army of an existing government.[77] As Malcolm X said of Dr. Martin Luther King, Jr.'s March on Washington, "Who ever heard of angry revolutionists swinging their bare feet together with their oppressor in lily-pad park pools, with gospels and guitars and 'I have a dream' speeches?"[78] But the human self-domestication hypothesis predicts that the more peaceful strategy will be more effective. Violent protests will only increase the perception of threat, setting off the feedback loop of reciprocal dehumanization. Violence will only escalate as people on the extreme end of any political ideology will dehumanize those they perceive threatening their group.

The political scientist Erica Chenoweth initially believed "that power flows from the barrel of a gun . . . although it was tragic, it was logical for people . . . to use violence to seek their change."[79] Chenoweth predicted that peaceful resistance, including protests, boycotts, and strikes, might work for "softer rights" like environmental reforms, gender rights, or labor reforms, "but it can't work generally if you're trying to overthrow a dictator, or become a new country."[79]

To test her hypothesis, Chenoweth collected data on all major peaceful and violent campaigns since 1900 that had tried to achieve the difficult goal of regime change. To her surprise, peaceful campaigns had proved twice as likely to succeed, and violent resistance four times more likely to fail.

Success Rates of Nonviolent and Violent Campaigns, 1900–2006

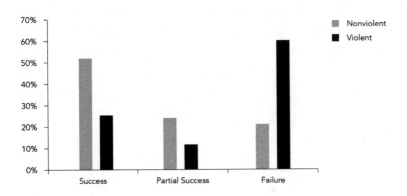

Once success had been achieved, nonviolent campaigns were more likely to establish a democracy that was less likely to relapse into a civil war.[80] And this trend is increasing over time, with peaceful campaigns increasingly likely to succeed.

Chenoweth attributes the success of peaceful campaigns to the sheer number of people who can be involved. On average, peaceful campaigns involve 150,000 more people than violent campaigns. Women, children, the elderly can all participate in peaceful protests. And while violent campaigns tend to take place covertly, under the cover of darkness, peaceful campaigns can be conducted in the open, for all to see.[80]

All protests must balance the tension between gaining attention for their cause and recruiting support. One study found that extreme protest tactics, like roadblocks, destroying property, and interpersonal violence, although good for gaining media attention and publicity, actually decrease popular support for the movement.[81]

In contrast, watching thousands, sometimes millions, of peaceful demonstrators, including women and children, singing songs and chanting peacefully, works to decrease the perception of the movement as a threat. Chenoweth found that security forces during peaceful protests are more likely to defect than they are during violent protests.

Success Rates by Decade, 1940–2006

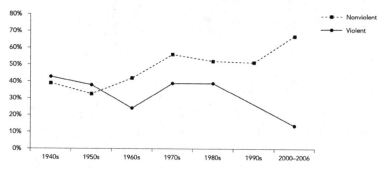

Data from Chenoweth on the success rates of nonviolent versus violent protests.

In 2017, the American Civil Liberties Union, which defended the right of white supremacists to protest, amended their policy after the violence in Charlottesville that ended with a thirty-two-year-old woman, Heather Heyer, killed by a car attack. The ACLU supports the rights of all groups to protest—but it must be peaceful protest. They no longer support protesters—from any side—who are armed.

The right to assemble is fundamental to democracy. But to those who seek change, it is the "peaceful" part of assembly that will reduce the possibility that someone will perceive you as a threat. Friendliness wins. Your peaceful effort is more likely to enact lasting change.

In 2017, almost monthly protests erupted at U.C. Berkeley, the birthplace of the 1960s Free Speech Movement, over whether far-right conservatives like Milo Yiannopoulos and Ann Coulter should be allowed to speak on campus.

Liberals called the rhetoric of white supremacists, neo-Nazis, and the alt-right "hate speech" and called for them to be banned from campus. In response, the alt-right invoked the First Amendment of the Constitution, which says "Congress shall make no law . . . abridging the freedom of speech,"[82] and claimed that "hate speech" was code for censorship.

The legal framework around attempts to prohibit hate speech is complicated because hate speech has been so hard to define. Is it "any form of expression deemed offensive to any racial, religious, ethnic or national group"[83]? Or is it only hate speech when it is "directed at members of a historically subordinated group"[84]?

In the United States, there are some limits on what can be said. You cannot slander someone, or shout obscenities on national television. You cannot threaten someone with violence. You cannot write a manual of how to be a hit man (like the one published, then prosecuted, in 1997). But as long as you avoid the few legal exceptions, your speech is protected.

Other democratic countries have laws against hate speech. In Australia, it is against the law to "offend, insult, humiliate or intimidate another person or a group of people."[85] In 2000, a website denying the Holocaust was found to be in violation of Australian law. In Germany, hate speech is defined as that which "incites hatred against a national, racial, religious group."[86] In 2017, German law directed social media sites like Facebook to remove hate speech within twenty-four hours or face a 50-million-euro fine. In Israel, hate speech includes expression "offending religious sensitivities."[87]

The human self-domestication hypothesis makes a clear prediction: When members of one group dehumanize, or refer to people in another group as less than human, they prime anyone listening for the worst acts of violence. It follows that the most dangerous form of hate speech would liken people to animals, or machines, or use words to describe others that elicit visceral emotions of disgust like "garbage," "parasites," "body fluids," or "filth."

Recall from chapter 8 that we know from fMRI studies that areas of the brain associated with the theory of mind network show less activity when we dehumanize other people.[88] Just overhearing someone else dehumanizing other people makes it more likely that you will dehumanize members of that group. This effect is present even in children.

Without threatening the freedom of speech, we can promote strong cultural norms against dehumanizing language.

When someone on TV or in a newspaper or any medium refers to a person or group as less than human, alarm bells should start ringing. As citizens, we can make sure such language is never normalized. As the Italian poet Giambattista Basile wrote, "though the tongue has no bones, it can break a spine."[89]

"Can you imagine—can you imagine these people, these animals over in the Middle East, that chop off heads, sitting around talking and seeing that we're having a problem with waterboarding?" Donald Trump said on his campaign trail. "We should go for waterboarding and we should go tougher than waterboarding."[90]

The presidential campaign of Donald Trump was unique for many reasons, but one of the most disturbing was the dehumanizing rhetoric he used throughout the campaign. Trump had an uncanny intuition for groups his constituents would consider outsiders and was adept at framing these outsiders as threatening. Trump called reporters who insulted his supporters "scum," "slime," and "disgusting." He called Hillary Clinton "nasty" and her supporters "animals."[91]

After generating a list of outsiders and emphasizing the threat they posed, Trump went on to encourage violence against them. He advocated torture, the death penalty, and deportation for refugees from war-torn countries. Journalists were not safe at his campaign events and had to be contained in pens for their own protection. Even his rhetoric was riddled with violence. He said he wanted to "punch [a protester] in the face," was pleased that a protester was "roughed up,"[92] and boasted that he could stand in the middle of Fifth Avenue and "shoot someone, and I wouldn't lose any voters."[93]

The political system of the United States is based on the democratic principle that every person, even your worst enemy, deserves to be counted as human. We need to work together as a society to shun leaders who dehumanize others and encourage those who, regardless of political party, insist on the humanity of others.

URBAN ANIMALS

Self-domestication's most powerful consequence was that it allowed us to live in larger, denser groups. While Neanderthals in the Upper Paleolithic could likely support bands of only a dozen or more people, we could support semipermanent settlements with populations in the hundreds. Eventually settlements became more permanent, and populations grew from hundreds to thousands to millions.

In 2008, we became an urban species. More people now live in cities than in rural areas. In many ways, this is good news. Even in the poorest countries, life in cities is, by some measures, preferable to that in rural areas. There is a better chance for upward mobility, education, and decent living standards.

At their best, cities create vibrant communities that mix people from different countries, ethnicities, sexual orientations, and races. This diversity facilitates contact, which in turn promotes tolerance, driving innovation and economic growth. Fortunately, we already know how to build cities that promote contact, because architecture, like all technology, is an extension of ourselves. What seems to work best are mid-rise buildings (twelve stories seems to be the upper limit), where mothers can keep an eye on children playing outside and residents can watch people passing below; mixed-income housing, where people from different professions and socio-economic status can live side by side; sidewalks that lead to small businesses, cafés, and restaurants; people working in these businesses who know their customers; gardens and playgrounds where mothers can talk and their children can make friends.[94–96]

In the 1950s, the West Village, New York, was this kind of city. The urbanist Jane Jacobs described the intricate "ballet" outside her apartment every morning. "The strangers on Hudson Street, the allies whose eyes help us natives keep the peace of the street, are so many that they always seem to be different people. . . . When you see the same stranger three or four times on Hudson Street, you begin to nod."[94]

At their worst, cities prevent social contact. High-rises create neighborhoods in which you can live on the same floor with someone for years and never meet, neighborhoods with no sidewalks and only big-box chain stores and fast-food restaurants, with gates and fences that prevent you from leaving or wandering around communities, with highways that cut through communities with no crosswalks or green spaces.

Many of our cities are racially segregated, a segregation that started just after World War II. The government poured money into highways that led to the suburbs, facilitating the "white flight" out of city centers. Government-sanctioned racial covenants prevented black people from buying houses in these suburbs. The Federal Housing Authority denied mortgages to people based on race and went even further, "redlining" entire suburbs. This physical distance between black and white communities destroyed opportunities for contact and made it easy for each community to dehumanize the other.

Urban architecture can also design certain people out of certain spaces. "Hostile architecture" is the term for the sloped windowsills and grated metal on steps that keep people from sitting on them, or for the balls on ledges and uneven pavements meant to deter skateboarders. The most hostile architecture of all is reserved for those most vulnerable to dehumanization, homeless people, who have to contend with spikes built under bridges and the awnings of buildings that might offer them shelter, armrests between park benches where they might have slept, and sprinklers meant to chase them from an otherwise welcoming green space.

The rest of us are not immune to the effects of this kind of architecture. According to the writer Alex Andreou, "By making the city less accepting of the human frame, we make it less welcoming to all humans. By making our environment more hostile, we become more hostile within it."[97]

Cities need to be designed to create conditions similar to those that were so beneficial to our ancestors. "Cities need to promote contact. And to do this they need institutional support," says the urban planner Mai Nyugen. Nyugen suggests subsidized housing in the cities, or close to transit lines, that can efficiently move people to job centers. "Exposure creates tolerance," she points out.

Cities should be places where people from different backgrounds, perspectives, and lived experiences can freely mix and exchange ideas. For our ancestors, these were settlements along trade routes, where far-flung travelers could share ideas, technology, and merchandise. For us, these are common areas—parks, cafés, theaters, restaurants—where we can meet and become familiar with neighborhood faces.

Our habitat has changed, but we have not. We are at our most productive when we live in large, cooperative groups. We are at our most innovative when we exchange ideas with people from diverse backgrounds—even those with whom we vehemently disagree. We are at our most tolerant when the architecture of our society facilitates tolerance. In order to maintain a healthy democracy and express the best of human nature, we need to design the spaces we live in so we can meet each other without being afraid, disagree without being disagreeable, and make friends with those least like ourselves.

9

Circle of Friends

Claudine André stormed up the stairs of the shelled-out building requisitioned by the army. It was the Democratic Republic of Congo's second war. The Rwandans had the city surrounded and Kinshasa had been under siege for a month. There was no food. No running water. On the outskirts of the city, Hutu soldiers threw tires on Tutsis and set them alight.

Claudine's husband was half Tutsi and half Italian, and he had been hiding in the Italian embassy for weeks. She went back and forth from her home to the embassy, bringing news and supplies. It was not safe for him to go outside, and even if it had been, there was nowhere to go. There were no planes and no helicopters. Everyone who could leave had already left.

When Claudine reached the second floor of the makeshift army office, she was stopped by soldiers.

"I would like to see the general."

"He is busy."

"I will wait."

Claudine waited, but not as long as she might have. The general, informed that a white woman with blazing red hair was standing outside his office, was unable to contain his curiosity.

"What can I do for you, Madame?"

"Your soldiers are cutting down the trees in the park."

"Yes?"

"I have twelve bonobos in my care. They have been orphaned by the war, and they will need somewhere to live when the war is over."

The twelve bonobos were sleeping in Claudine's garage. Every day, they piled into her SUV and she drove them to a small forest at the back of a school. The park Claudine was referring to had been one of the private retreats of the former dictator, Mobutu Sese Seko, a set of sprawling gardens he had filled with tropical plants and animals. But the dictator had died long ago. The soldiers were in charge now.

"Bonobos are the pride of Congo. They live nowhere else. This park should be for them."

A bomb fell close enough to the building that the walls shook and plaster fell from the ceiling. Claudine continued calmly.

"Please tell your soldiers to stop cutting down the trees."

"Madame, you must leave. It is not safe—"

Another bomb fell.

"These bonobos need protection."

Perhaps realizing that the woman would not leave until he agreed, the general said, "I will tell them."

Claudine stood patiently. Another bomb.

"I hereby make you guardian of the park! File a report with me every six months. The men will be informed. Now, Madame, please!"

You might question Claudine's sanity, arguing over trees for bonobos in the middle of a war. But Claudine loves animals. Any creature who is sick or hurt can depend on her for care. During the First Congo War, she brought food for the starving animals in the zoo. At one point, besides sixty-three bonobos, she was looking after three African gray parrots, a galago, three dogs, ten cats, and a putty-nosed monkey.

Claudine started dozens of Kindness Clubs around Kinshasa where children learned that animals had thoughts and feelings and deserved to be treated with compassion. One morning, Claudine was talking to a Kindness Club visiting the sanctuary when a man stood up and interrupted her.

"How can you talk about animals like this?" he asked her.

"People in Congo are suffering. These bonobos have more food and better care than these children in front of you."

"I teach children to be kind to animals," Claudine replied, "so they will be kind to one another."

THE ANIMAL CONNECTION

Does kindness toward animals really translate into kindness toward others? If anything, researchers have traditionally argued that relating to animals creates stress because it challenges our belief that we are special and different.[1] According to this view, we do not like to be reminded that we share part of our nature with animals, which is what makes comparing people to animals such an effective dehumanizing tactic.

Students were asked which of fifteen different factors were most likely to drive prejudice and dehumanization. The majority blamed ignorance, closed-mindedness, the media, parental influence, and cultural differences. In contrast, they saw their perception of animals as irrelevant, even though they acknowledged that dehumanization was a process of seeing other people as more animal-like. These same students saw education and intergroup contact as the main solutions to improving intergroup relations.[2]

If there is an essential lesson to be learned from psychology research, it is that we are not always aware of what shapes our attitudes and behavior. People unconsciously judge one another based on a variety of physical traits, and they do the same with animals. My collaborator Margaret Gruen showed photos of a variety of dog breeds to veterinarians and the general public. When we asked people from both groups to rate how sensitive each dog is to pain, even though there is no scientific evidence that any dogs experience pain differently from others, the general public consistently scored smaller dogs as being more sensitive to pain than larger dogs. Veterinarians also scored different breeds as differing in pain sensitivity, even though no veterinary school teaches that such differences exist. Both groups also rated dogs with reputations for aggression as less sensitive to pain. They even rated darker-colored dogs as experiencing less pain than lighter-colored dogs of the same breed.

Neither are we entirely aware of how our relationship with animals relates to how we view one another. For example, although we may not think kindness toward animals relates to kindness toward people, we generally believe that cruelty toward animals predicts cruelty toward people.

We know cruelty toward animals during childhood is often a warning sign of more dangerous behavior to come. It is one of the childhood symptoms of psychopathy. This relationship is not just apparent in extreme forms of mental illness. Attitudes toward animals are also correlated with the general public's attitudes toward other people. The psychologists Gordon Hodson and Kristof Dhont examined whether people who think humans are superior to animals are also more likely to rate some human groups as superior to other groups. They found that "seeing humans as different from and superior to animals plays a key role in animalistically dehumanizing human outgroups including immigrants, black people, or ethnic minorities."[3]

In another study,[4] Hodson looked specifically at the width of the animal-human divide, asking people how much they agreed with statements like "Humans are not the only creatures who have thoughts; some nonhuman animals can think too." People who saw a greater difference between animals and humans were more likely to dehumanize immigrants and to agree with statements like "Immigrants are getting too demanding in their push for equal rights."[4] On the other hand, people who believed animals were more similar to humans were less likely to dehumanize immigrants. The animal-human divide, or the perceived distance between people and animals, seems integrally related to the perceived distance between groups of people.

DINGOES ARE OUR MOTHERS

In the Western industrialized world, the divide between humans and dogs has diminished dramatically in the last few decades. Dogs have shifted from being viewed primarily as work animals or status symbols to being regarded as full-fledged family members. The love lavished on pet dogs might seem like just another excess of modern life, but prehistoric graves suggest that this love is more ancient. Burials from more than 10,000 years ago have been found on several continents where the dead are laid to rest cradling a dog in their arms.

In one of the most egalitarian cultures in the world, the love between humans and dogs is even more remarkable. The Martu Aboriginals live in a remote region in western Australia, one of the most spectacularly hostile places on earth. They are the traditional landowners of an area the size of Connecticut that extends from the Great Sandy Desert to Wiluna. The scrubby vegetation, scarcity of water, and burning sun prompted the first Dutch explorers to proclaim the land uninhabitable.

The Martu, however, are part of a vast network of Aboriginals who have lived in this area for thousands of years.[5] They are one of the few hunter-gatherer bands left and were among the last Aboriginals to make contact with Europeans, as late as the 1960s. They pass down secrets for finding water and mapping the terrain to successive generations through complex artwork that has become famous around the world. The Martu, like all Australian Aboriginals, have a deep and complex relationship with the land and all the animals that live on it. This spiritual tradition is called the Dreamtime. A prominent figure in the Dreamtime is the dingo.

Like all dogs, dingoes originated from a wolflike ancestor that, in this case, arrived in Australia from Asia at least 5,000 years ago. But dingoes evolved without the radical, human-controlled breeding that created most modern dog breeds. Dingoes walk between the worlds of tame and wild, some living with people, others living far away in hostile wilderness. Unlike feral pet dogs, these wild dingoes thrive without us, but they can be close to people. "These dingoes are our mothers," one of the Martu explained to the anthropologist Doug Bird when he interviewed them.

This is not just a metaphor. They explained to him that when they were children, Martu families left the camp to hunt and forage, and older children watched the younger ones and took them home when they were tired. When the children arrived back in camp, dingoes, who traveled with them, regurgitated their food, just as they would for their puppies, and the older children cooked the protein-rich mash over the fire, tiding the children over until their parents brought back to camp meat, roots, nuts, or berries for everyone. Waiting for their parents to return, the children curled up beside the dingoes for warmth, knowing that nothing would harm them while the dingoes were on watch.

Dogs have likely been part of the human family in the outback for thousands of years. It's an extraordinary relationship, not just on the part of the dingoes, mothering and caring for a species that now more often hunts and persecutes them, but also on the part of the Australian Aboriginals. This highly egalitarian society did not see the dingoes as pests or workers, but as family. Something must have changed in our relationship with dogs that pushed dogs out of the family circle—probably around the time of industrialization.

European dog breeds have a surprisingly recent origin.[6] People cultivated them during the Victorian age, when the physical appearance of a dog became more important than the job the dog was supposed to perform. Before the Victorians, any large dog was called a mastiff, any dog who hunted hares was a harrier, and any lapdog was a spaniel.[7]

The first dog shows were held in the late nineteenth century and were created to select dogs with "superior" traits that could further improve "pure" bloodlines. Award-winning dogs brought their owners prestige and significant income. Dogs became a commodity to be traded, and each breed came with a narrative about what made it superior—especially compared to dogs without a known pedigree. As the dog breeder and author Gordon Stables wrote in 1896, "Nobody who is anybody can afford to be followed about by a mongrel dog."[8]

Groups of dogs began to be perceived as either superior or inferior. Owning the offspring of a winning pedigreed dog or a dog of a fashionable breed quickly became a signal of social status. Well-bred dogs came to represent power and high rank.[8] "Breed and breeds were examples of order and hierarchy, and of tradition, even though much, if not all, were invented."[8] European dog breeds were the product of a culture that was obsessed with class and hierarchy, so much so that it gave birth to the eugenics movement.

Remember that Social Dominance Orientation (SDO) is a measure of belief in group hierarchy. Strongly agreeing with the statement "Some groups are superior to others" indicates that a person has high SDO. Together with my graduate student Wen Zhou, we created the dog SDO survey, in which we simply replaced the word "group" with the word "breed." We asked more than a thousand people whether they agreed or disagreed with statements like "Some breeds of dogs are superior to other breeds" and "We don't need to guarantee that every breed of dog has the same quality of life."

Many respondents strongly agreed with these hierarchical statements. These same people also had a stronger preference for purebred dogs. But perhaps most remarkable was their response to the original human SDO survey questions. We found that the same people who perceive distinct group hierarchy between dog breeds also perceive distinct hierarchy between human groups. The tendency to have a strong orientation in favor of dog SDO paralleled the orientation in favor of human SDO. We also found that dog owners have slightly higher SDO ratings than non-dog-owners, although people who are bonded to their dog and view them as family have significantly lower SDO than the average person.

More egalitarian hunter-gatherers most likely accepted dogs into their families, as did the Martu Aboriginals. It was during agriculture and industrialization that dogs likely went from family members to workers to status symbols that reinforced social hierarchy. As liberal democracies and economic prosperity spread, our dogs rapidly returned to their place in our families. Our egalitarian attitudes toward one another are reflected in how we view and treat our nonhuman best friends.

Our views on the treatment of dogs might also reflect what we view as acceptable treatment of others—other human groups and species. High dog SDO is associated with a readiness to see "inferior" human group members as animals.

In searching for ways to bridge the difference between ourselves and those we might think of as outsiders—both human and animal—our friendship with dogs might be the most powerful and the most accessible. No one who has ever loved a dog would question a dog's ability to think, suffer, or love. No one who has ever had the gift of being loved by a dog would think that their love was worth less. Friendship is the world's greatest equalizer. No one could have predicted how important dogs would be to us. They evolved from fanged carnivores when we were the superpredators of the Paleolithic. Instead of using the fear and aggression that had made them so successful, they gently came closer. It took many generations, but we found enough common ground to become important to each other. Two legs or four, dark or light, it makes no difference to how much they love us. And that love can change our lives. At least, it changed mine.

Ever since we thought up clever weapons to outcompete the other, stronger human species, we have put undue emphasis on intelligence. We have used our ideas about intelligence to create a linear scale that allows us to subject both animals and people to cruelty and suffering. My own dog, Oreo, taught me that every living creature has its own variety of genius, and everyone is born into this world with a mind brilliantly capable of solving the problems relevant to their survival.

Discovering the genius in Oreo opened my eyes to the mental potential of other animals. Because of Oreo, I looked a little closer at the capabilities of chimpanzees. Because of Oreo, I turned to bonobos and discovered animals who see every stranger as a potential friend.

Loving Oreo led me to the most valuable lesson of all: Our lives should be measured not by how many enemies we have conquered, but by how many friends we have made. That is the secret to our survival.

Acknowledgments

We had a first draft ready around October 2016. That draft ended with a warning that the very worst of our nature could resurface and express itself in any place or culture. But after the U.S. elections, we threw away half or more of that first draft. We realized we had to offer solutions. Real solutions required an even deeper dive into the scientific literature. We had to become experts in areas of social psychology, history, and political science with which we had been unfamiliar. We spent two more years researching, restructuring, and rewriting the book so that we could not only showcase our species' unusual friendliness but also offer help in thinking about the root causes and solutions of our most vexing problems.

Disagreement and debate in science are healthy and exciting. Disagreement often drives research that leads to advances in our understanding. Scientists rely on skepticism and empirical debate as a road to truth. Although we have done our best to represent the literature fairly, not every scientist will agree with everything we report. Unlike the scientific papers we write, to assure ease of reading, we did not always highlight alternative perspectives or competing data in the main text. However, we have provided extensive references and notes that cover important details and alternative findings. If you are interested, there are many ways to get access to these sources and read them for yourself.*

The book benefited immensely from the efforts of many people who took the time to help us along the journey. The biggest thank-you must go to our editor, Hilary Redmon, who for five years has been the kindest, most thoughtful editor two writers could wish for. If this were an academic publication, she would be listed as a co-author. She went above and beyond typical editorial duties to help us dramatically improve the organization and writing of this manuscript. We are deeply indebted to her, and it was a joy to work together.

Our agent, Max Brockman, was instrumental in shaping our book proposal as we initially began brainstorming about what the book might be. His encouragement was critical as we decided to take on this project.

There would be no self-domestication hypothesis without Richard Wrangham and Mike Tomasello. Our collaboration and their own work inspired many of the ideas in this book. They also tirelessly endured endless discussions that

---

* Most of the research we review is available to you online. Google has a function called Google Scholar that allows you to download many of the papers; many scientific journals allow free access to their papers through their websites; you can search the website of the scientists who authored the papers and download their papers for free; and finally, there is nothing scientists like more than sharing their papers. If you write to them, they may happily share papers we discuss in this book if you cannot get access to them otherwise.

helped us synthesize so many disparate ideas. Our colleague Walter Sinnott-Armstrong was also a fantastic mentor—always ready to provide friendly encouragement and helpful arguments along the way.

Thank you to our many Duke students who have shaped much of our thinking as we grappled with the primary literature in seminar classes and lab meetings. In particular, we must thank graduate students and postdocs Victoria Wobber, Alexandra Rosati, Evan MacLean, Jingzhi Tan, Kara Walker, Chris Krupenye, Aleah Bowie, Wen Zhou, Margaret Gruen, and Hannah Salomons for all the stimulating conversations that in large part motivated this work. You were all extremely patient as you tolerated our split attention while we worked on this book. Aleah and Wen were especially generous in helping us as we struggled to understand the vast social psychology literature relevant to the best and worst of human nature.

Thank you to our department staff and lab coordinators Lisa Jones, Ben Allen, James Brooks, Kyle Smith, Maggie Bunzey, Morgan Ferrans, and Madison Moore who kept everything going as we were writing. Finally, thanks to Jessica Tan and Rong Xiang for creating many of the figures included in the text.

Our own research, reported on throughout the book, was supported by federal and foundation funds including the Office of Naval Research (NOOO14-12-1-0095, NOOO14-16-1-2682), the Eunice Kennedy Shriver National Institute of Child Health and Human Development (NIH-HD070649; NIH-1R01HD097732), the National Science Foundation

(NSF-BCS-08-27552), the Stanton Foundation, the Canine Health Foundation, and the Templeton World Charity Foundation.

Loving thanks to Vanessa's mom, "Bobo" (Jacquie Leong), who helped wrangle kids at various stages, and to Brian's folks, "Mema" and "Pops" (Alice and Bill Hare), who started it all by bringing home a wriggling black Labrador called Oreo. Thank you to our friends, who put up with various stages of ranting and raging and occasional public break-downs and who so quickly learned not to ask "Is the book done yet?" Thank you to our dogs, Tassie and Congo, who loved us and reminded us every day how friendliness is the winning strategy. Finally, thank you to our sweet children, Malou and Luke. Yes, the book is finally done. Yes, we can play with you now.

We hope our book will inspire all humans to show even more compassion toward one another and the animals with which we share this earth. If you wish to help Claudine André with her mission to save bonobos and to help encourage the youth of Congo to show kindness to humans and all animals—including dogs—please consider donating to Friends of Bono-bos: friendsofbonobos.org. If you would like to follow our research group's progress, you can visit dukedogs.com.

Notes

Introduction

1. Elliot Aronson, Shelley Patnoe, *Cooperation in the Classroom: The Jigsaw Method* (London: Pinter & Martin, 2011).

2. D. W. Johnson, G. Maruyama, R. Johnson, D. Nelson, L. Skon, "Effects of Cooperative, Competitive, and Individualistic Goal Structures on Achievement: A Meta-Analysis," *Psychological Bulletin* 89, 47 (1981).

3. D. W. Johnson, R. T. Johnson, "An Educational Psychology Success Story: Social Interdependence Theory and Cooperative Learning," *Educational Researcher* 38, 365–79 (2009).

4. M. J. Van Ryzin, C. J. Roseth, "Effects of Cooperative Learning on Peer Relations, Empathy, and Bullying in Middle School," *Aggressive Behavior* (2019).

5. C. J. Roseth, Y.-k. Lee, W. A. Saltarelli, "Reconsidering Jigsaw Social Psychology: Longitudinal Effects on Social Interdependence, Sociocognitive Conflict Regulation, Motivation, and Achievement," *Journal of Educational Psychology* 111, 149 (2019).

6. Charles Darwin, *Descent of Man, and Selection in Relation to Sex,* new edition, revised and augmented (Princeton, New Jersey: Princeton University Press 1981; Photocopy of original London: Murray Publishing 1871).

7. Brian Hare, "Survival of the Friendliest: *Homo sapiens* Evolved via Selection for Prosociality," *Annual Review of Psychology* 68, 155–86 (2017).

Domestication over the generations does not, as was once thought, decrease intelligence; it increases friendliness. When a species of animal is domesticated, it undergoes many changes that appear completely unrelated to one another. This pattern of changes—called the domestication syndrome—can show up in the shape of the face, the size of the teeth, and the pigmentation of different body parts or hair, and it can include changes to hormones, reproductive cycles, and the nervous system. What we discovered in our research is that it also can increase a species' ability to coordinate and communicate with others.

All these seemingly random changes are tied to development. The brains and bodies of domesticated species develop differently than those of the less friendly species from which they evolved. The behaviors that facilitate social bonding, such as play, appear earlier and are retained longer—often through adulthood—in domesticated species than in other closely related species.

Studying domestication in other species has allowed us to understand how our own cognitive superpower probably evolved.

8. R. Kurzban, M. N. Burton-Chellew, S. A. West, "The Evolution of Altruism in Humans," *Annual Review of Psychology* 66, 575–99 (2015).

9. Frans de Waal, *Peacemaking Among Primates* (Cambridge, MA: Harvard University Press, 1989).

10. R. M. Sapolsky, "The Influence of Social Hierarchy on Primate Health," *Science* 308, 648–52 (2005).

11. N. Snyder-Mackler, J. Sanz, J. N. Kohn, J. F. Brinkworth, S. Morrow, A. O. Shaver, J.-C. Grenier, R. Pique-Regi, Z. P. Johnson, M. E. Wilson, "Social Status Alters Immune Regulation and Response to Infection in Macaques," *Science* 354, 1041–45 (2016).

12. C. Drews, "Contexts and Patterns of Injuries in Free-Ranging Male Baboons (*Papio cynocephalus*)," *Behaviour* 133, 443–74 (1996).

13. M. L. Wilson, C. Boesch, B. Fruth, T. Furuichi, I. C. Gilby, C. Hashimoto, C. L. Hobaiter, G. Hohmann, N. Itoh, K.J.N. Koops, "Lethal Aggression in Pan Is Better Explained by Adaptive Strategies than Human Impacts," *Nature* 513, 414 (2014).

14. Thomas Hobbes, *Leviathan* (London: A&C Black, 2006).

15. Frans de Waal, *Chimpanzee Politics: Power and Sex Among Apes* (Baltimore: Johns Hopkins University Press, 2007).

16. L. R. Gesquiere, N. H. Learn, M. C. M. Simao, P. O. Onyango,

S. C. Alberts, J. Altmann, "Life at the Top: Rank and Stress in Wild Male Baboons," *Science* 333, 357–60 (2011).

17. M. W. Gray, "Mitochondrial Evolution," *Cold Spring Harbor Perspectives in Biology* 4, a011403 (2012), published online September 1, 10:1101/cshperspect.a011403.

18. L. A. David, C. F. Maurice, R. N. Carmody, D. B. Gootenberg, J. E. Button, B. E. Wolfe, A. V. Ling, A. S. Devlin, Y. Varma, M. A. Fischbach, S. B. Biddinger, R. J. Dutton, P. J. Turnbaugh, "Diet Rapidly and Reproducibly Alters the Human Gut Microbiome," *Nature* 505, 559–63 (2014), published online EpubJan 23, 10:1038/nature12820.

19. S. Hu, D. L. Dilcher, D. M. Jarzen, "Early Steps of Angiosperm-Pollinator Coevolution," *Proceedings of the National Academy of Sciences* 105, 240–45 (2008).

20. B. Hölldobler, E. O. Wilson, *The Superorganism: The Beauty, Elegance, and Strangeness of Insect Societies* (New York: W. W. Norton, 2009).

21. B. Wood, E. K. Boyle, "Hominin Taxic Diversity: Fact or Fantasy?" *American Journal of Physical Anthropology* 159, 37–78 (2016).

22. A. Powell, S. Shennan, M. G. Thomas, "Late Pleistocene Demography and the Appearance of Modern Human Behavior," *Science* 324, 1298–1301 (2009).

23. Steven E. Churchill, *Thin on the Ground: Neandertal Biology, Archeology, and Ecology* (Hoboken, NJ: John Wiley & Sons, 2014), vol. 10.

24. A. S. Brooks, J. E. Yellen, R. Potts, A. K. Behrensmeyer, A. L. Deino, D. E. Leslie, S. H. Ambrose, J. R. Ferguson, F. d'Errico, A.M.J.S. Zipkin, "Long-Distance Stone Transport and Pigment Use in the Earliest Middle Stone Age," *Science* 360, 90–94 (2018).

25. N. T. Boaz, *Dragon Bone Hill: An Ice-Age Saga of Homo Erectus,* edited by R. L. Ciochon (Oxford and New York: Oxford University Press, 2004).

26. C. Shipton, M. D. Petraglia, "Inter-continental Variation in Acheulean Bifaces," in *Asian Paleoanthropology* (New York: Springer, 2011), 49–55.

27. W. Amos, J. I. Hoffman, "Evidence That Two Main Bottleneck Events Shaped Modern Human Genetic Diversity," *Proceedings of the Royal Society B: Biological Sciences* (2009).

28. A. Manica, W. Amos, F. Balloux, T. Hanihara, "The Effect of

Ancient Population Bottlenecks on Human Phenotypic Variation," *Nature* 448, 346–48 (2007).

29. S. H. Ambrose, "Late Pleistocene Human Population Bottlenecks, Volcanic Winter, and Differentiation of Modern Humans," *Journal of Human Evolution* 34, 623–51 (1998), published online Epub1998/06/01/.

30. J. Krause, C. Lalueza-Fox, L. Orlando, W. Enard, R. E. Green, H. A. Burbano, J.-J. Hublin, C. Hänni, J. Fortea, M. De La Rasilla, "The Derived FOXP2 Variant of Modern Humans Was Shared with Neandertals," *Current Biology* 17, 1908–12 (2007).

31. F. Schrenk, S. Müller, C. Hemm, P. G. Jestice, *The Neanderthals* (Routledge, 2009).

32. S. E. Churchill, J. A. Rhodes, "The Evolution of the Human Capacity for 'Killing at a Distance': The Human Fossil Evidence for the Evolution of Projectile Weaponry," in *The Evolution of Hominin Diets* (New York: Springer, 2009), 201–10.

33. B. Davies, S. H. Bickler, "Sailing the Simulated Seas: A New Simulation for Evaluating Prehistoric Seafaring" in *Across Space and Time: Papers from the 41st Conference on Computer Applications and Quantitative Methods in Archaeology, Perth, 25–8 March 2013* (Amsterdam: Amsterdam University Press, 2015), 215–23.

34. O. Soffer, "Recovering Perishable Technologies Through Use Wear on Tools: Preliminary Evidence for Upper Paleolithic Weaving and Net Making," *Current Anthropology* 45, 407–13 (2004).

35. J. F. Hoffecker, "Innovation and Technological Knowledge in the Upper Paleolithic of Northern Eurasia," *Evolutionary Anthropology: Issues, News, and Reviews* 14, 186–98 (2005).

36. O. Bar-Yosef, "The Upper Paleolithic Revolution," *Annual Review of Anthropology* 31, 363–93 (2002).

37. S. McBrearty, A. S. Brooks, "The Revolution That Wasn't: A New Interpretation of the Origin of Modern Human Behavior," *Journal of Human Evolution* 39, 453–563 (2000).

38. M. Vanhaeren, F. d'Errico, C. Stringer, S. L. James, J. A. Todd, H. K. Mienis, "Middle Paleolithic Shell Beads in Israel and Algeria," *Science* 312, 1785–88 (2006).

39. G. Curtis, *The Cave Painters: Probing the Mysteries of the World's First Artists* (New York: Anchor, 2007).

40. H. Valladas, J. Clottes, J.-M. Geneste, M. A. Garcia, M. Arnold, H. Cachier, N. Tisnérat-Laborde, "Palaeolithic Paintings: Evolution of Prehistoric Cave Art," *Nature* 413, 479 (2001).

41. N. McCarty, K. T. Poole, H. Rosenthal, *Polarized America: The Dance of Ideology and Unequal Riches* (Cambridge, MA: MIT Press, 2016).

42. Charles Gibson, "Restoring Comity to Congress," paper presented at the Shorenstein Center on Media, Politics and Public Policy, Harvard Kennedy School, January 1, 2011, https://shoren steincenter.org/restoring-comity-to-congress/.

43. C. News, in *CBS News* (2010).

44. John. A. Farrell. *Tip O'Neill and the Democratic Century: A Biography*. (New York: Little Brown, 2001).

45. R. Strahan, *Leading Representatives: The Agency of Leaders in the Politics of the U.S. House* (Baltimore: Johns Hopkins University Press, 2007).

46. J. Haidt, *The Righteous Mind: Why Good People are Divided by Politics and Religion* (New York: Vintage, 2012).

47. Politicians who live in Washington are criticized as "carpetbaggers" (C. Raasch, "Where do U.S. Senators 'live,' and does it matter?," *St. Louis Post-Dispatch*, September 2, 2016) and "out of touch" with their constituents (A. Delaney, "Living Large of Capital Hill," *Huffington Post*, July 31, 2012). But having members of Congress living in the same place would build relationships between them. As Ornstein said, "if you are standing on the sidelines of a soccer game with a colleague from across the aisle and his or her spouse, you're just going to have a harder time vilifying him as the incarnation of the devil when you get on the floor" (Gibson, "Restoring Comity to Congress"). Being part of the same community might restore a sense of shared experience and common purpose that our representatives have lost. If contact is not possible for our representatives, then at least there should be contact between the staffers. Few people know that our government is primarily run by idealistic young people in their twenties, hoping to make a difference for our democracy. The effectiveness of our representatives is linked to the quality of their staff (J. McCrain, "Legislative Staff and Policymaking," Emory University, 2018). They are the ones who provide their representatives with information, talking points, and briefings. When you call or visit your representative, a young staffer will answer or meet with you. When you hear a politician give a press conference, it was a staffer who briefed them. Staffers live in one of the most expensive areas in the United States and are paid almost nothing, some as little as $20,000 a year. When I spoke to one of my former students, who is a

staffer for a U.S. senator, he said there are very few opportunities for Republican and Democratic staffers to form any kind of relationship. He had never even been to lunch with a staffer from the other party. If the older generation is too entrenched in the current polarization, it would not take much to bring these more open-minded young people together. Many of these young people will be in politics for decades. Through friendships, even casual acquaintances, they will be able to accomplish what their more experienced elders have not. But by not giving them a chance, we are looking at even greater polarization in the future.

48. N. Gingrich, "Language: A Key Mechanism of Control," *Information Clearing House* (1996).

49. D. Corn, T. Murphy, "A Very Long List of Dumb and Awful Things Newt Gingrich Has Said and Done," in *Mother Jones* (2016).

50. S. M. Theriault, D.W.J.T.J.o.P. Rohde, "The Gingrich Senators and Party Polarization in the U.S. Senate," *The Journal of Politics* 73, 1011–24 (2011).

51. J. Biden, "Remarks: Joe Biden," National Constitution Center, 16 October 2017 (2017), published online https://constitutioncenter .org/liberty-medal/media-info/remarks-joe-biden.

52. S. A. Frisch, S. Q. Kelly, *Cheese Factories on the Moon: Why Earmarks Are Good for American Democracy* (Routledge, 2015).

53. Cass R. Sunstein, *Can It Happen Here?: Authoritarianism in America* (New York: Dey Street Books, 2018).

1 Thinking About Thinking

1. M. Tomasello, M. Carpenter, U. Liszkowski, "A New Look at Infant Pointing," *Child Development* 78, 705–22 (2007).

2. Brian Hare, "From Hominoid to Hominid Mind: What Changed and Why?" *Annual Review of Anthropology* 40, 293–309 (2011).

3. Michael Tomasello, *Becoming Human: A Theory of Ontogeny* (Cambridge, MA: Belknap Press of Harvard University Press, 2019).

4. Michael Tomasello, *Origins of Human Communication* (Cambridge, MA: MIT Press, 2010).

5. E. Herrmann, J. Call, M. V. Hernández-Lloreda, B. Hare, M. Tomasello, "Humans Have Evolved Specialized Skills of Social Cognition: The Cultural Intelligence Hypothesis," *Science* 317, 1360–66 (2007).

6. A. P. Melis, M. Tomasello, "Chimpanzees (*Pan troglodytes*) Coor-

dinate by Communicating in a Collaborative Problem-Solving Task," *Proceedings of the Royal Society B* 286, 20190408 (2019).

7. J. P. Scott, "The Social Behavior of Dogs and Wolves: An Illustration of Sociobiological Systematics," *Annals of the New York Academy of Sciences* 51, 1009–21 (1950).

8. Brian Hare and Vanessa Woods, *The Genius of Dogs* (Oneworld Publications, 2013).

9. B. Hare, M. Brown, C. Williamson, M. Tomasello, "The Domestication of Social Cognition in Dogs," *Science* 298, 1634–36 (2002).

10. B. Hare, M. Tomasello, "Human-like Social Skills in Dogs?" *Trends in Cognitive Sciences* 9, 439–44 (2005).

11. B. Agnetta, B. Hare, M. Tomasello, "Cues to Food Location That Domestic Dogs (*Canis familiaris*) of Different Ages Do and Do Not Use," *Animal Cognition* 3, 107–12 (2000).

12. J. W. Pilley, "Border Collie Comprehends Sentences Containing a Prepositional Object, Verb, and Direct Object," *Learning and Motivation* 44, 229–40 (2013).

13. J. Kaminski, J. Call, J. Fischer, "Word Learning in a Domestic Dog: Evidence for 'Fast Mapping,'" *Science* 304, 1682–83 (2004).

14. K. C. Kirchhofer, F. Zimmermann, J. Kaminski, M. Tomasello, "Dogs (*Canis familiaris*), but Not Chimpanzees (*Pan troglodytes*), Understand Imperative Pointing," *PLoS One* 7, e30913 (2012).

15. F. Kano, J. Call, "Great Apes Generate Goal-Based Action Predictions: An Eye-Tracking Study," *Psychological Science* 25, 1691–98 (2014).

16. E. L. MacLean, E. Herrmann, S. Suchindran, B. Hare, "Individual Differences in Cooperative Communicative Skills Are More Similar Between Dogs and Humans than Chimpanzees," *Animal Behaviour* 126, 41–51 (2017).

17. Jonathan B. Losos, *Improbable Destinies: Fate, Chance, and the Future of Evolution* (Penguin, 2017).

18. E. Axelsson, A. Ratnakumar, M.-L. Arendt, K. Maqbool, M. T. Webster, M. Perloski, O. Liberg, J. M. Arnemo, Å. Hedhammar, K. Lindblad-Toh, "The Genomic Signature of Dog Domestication Reveals Adaptation to a Starch-Rich Diet," *Nature* 495, 360–64 (2013).

19. G.-d. Wang, W. Zhai, H.-c. Yang, R.-x. Fan, X. Cao, L. Zhong, L. Wang, F. Liu, H. Wu, L.-g. Cheng, "The Genomics of Selection in Dogs and the Parallel Evolution Between Dogs and Humans," *Nature Communications* 4, 1860 (2013).

20. Y.-H. Liu, L. Wang, T. Xu, X. Guo, Y. Li, T.-T. Yin, H.-C. Yang,

H. Yang, A. C. Adeola, O. J Sanke, "Whole-Genome Sequencing of African Dogs Provides Insights into Adaptations Against Tropical Parasites," *Molecular Biology and Evolution* (2017).

2 The Power of Friendliness

1. S. Argutinskaya, in memory of D. K. Belyaev, "Dmitrii Konstantinovich Belyaev: A Book of Reminescences," edited by V. K. Shumnyi, P. M. Borodin, A. L. Markel, and S. V. Argutinskaya (Novosibirsk: Sib. Otd. Ros. Akad. Nauk, 2002), *Russian Journal of Genetics* 39, 842–43 (2003).
2. Brian Hare, Vanessa Woods, *The Genius of Dogs* (Oneworld Publications, 2013).
3. Lee A. Dugatkin, L. Trut, *How to Tame a Fox (and Build a Dog): Visionary Scientists and a Siberian Tale of Jump-Started Evolution* (Chicago: University of Chicago Press, 2017).
4. Darcy Morey, *Dogs: Domestication and the Development of a Social Bond* (Cambridge University Press, 2010); M. Geiger, A. Evin, M. R. Sánchez-Villagra, D. Gascho, C. Mainini, C. P. Zollikofer, "Neomorphosis and Heterochrony of Skull Shape in Dog Domestication," *Scientific Reports* 7, 13443 (2017).
5. E. Tchernov, L. K. Horwitz, "Body Size Diminution Under Domestication: Unconscious Selection in Primeval Domesticates," *Journal of Anthropological Archaeology* 10, 54–75 (1991).
6. L. Andersson, "Studying Phenotypic Evolution in Domestic Animals: A Walk in the Footsteps of Charles Darwin" in *Cold Spring Harbor Symposia on Quantitative Biology* (2010).
7. Helmut Hemmer, *Domestication: The Decline of Environmental Appreciation* (Cambridge: Cambridge University Press, 1990).
8. Jared Diamond, "Evolution, Consequences and Future of Plant and Animal Domestication," *Nature* 418, 700–707 (2002).
9. Jared Diamond, *Guns, Germs, and Steel: The Fates of Human Societies* (New York: W. W. Norton, 1999).
10. Lyudmila Trut, "Early Canid Domestication: The Farm-Fox Experiment; Foxes Bred for Tamability in a 40-year Experiment Exhibit Remarkable Transformations That Suggest an Interplay Between Behavioral Genetics and Development," *American Scientist* 87, 160–69 (1999).
11. M. Geiger, A. Evin, M. R. Sánchez-Villagra, D. Gascho,

C. Mainini, C. P. Zollikofer, "Neomorphosis and Heterochrony of Skull Shape in Dog Domestication," *Scientific Reports* 7, 13443 (2017).

12. L. Trut, I. Oskina, A. Kharlamova, "Animal Evolution During Domestication: The Domesticated Fox as a Model," *Bioessays* 31, 349–60 (2009).

13. A. V. Kukekova, L. N. Trut, K. Chase, A. V. Kharlamova; J. L. Johnson, S. V. Temnykh, I. N. Oskina, R. G. Gulevich, A. V. Vladimirova, S. Klebanov, "Mapping Loci for Fox Domestication: Deconstruction/Reconstruction of a Behavioral Phenotype," *Behavior Genetics* 41, 593–606 (2011).

14. A. V. Kukekova, J. L. Johnson, X. Xiang, S. Feng, S. Liu, H. M. Rando, A. V. Kharlamova, Y. Herbeck, N. A. Serdyukova, Z.J.N. Xiong, "Red Fox Genome Assembly Identifies Genomic Regions Associated with Tame and Aggressive Behaviours," *Evolution* 2, 1479 (2018).

15. E. Shuldiner, I. J. Koch, R. Y. Kartzinel, A. Hogan, L. Brubaker, S. Wanser, D. Stahler, C. D. Wynne, E. A. Ostrander, J. S. Sinsheimer, "Structural Variants in Genes Associated with Human Williams-Beuren Syndrome Underlie Stereotypical Hypersociability in Domestic Dogs," *Science Advances* 3 (2017).

16. L. A. Dugatkin, "The Silver Fox Domestication Experiment," *Evolution: Education and Outreach* 11, 16 (2018), published online Epub2018/12/07, 10:1186/s12052-018-0090-x.

17. B. Agnvall, J. Bélteky, R. Katajamaa, P. Jensen, "Is Evolution of Domestication Driven by Tameness? A Selective Review with Focus on Chickens," *Applied Animal Behaviour Science* (2017).

18. B. Hare, I. Plyusnina, N. Ignacio, O. Schepina, A. Stepika, R. Wrangham, L. Trut, "Social Cognitive Evolution in Captive Foxes Is a Correlated By-product of Experimental Domestication," *Current Biology* 15, 226–30 (2005).

19. B. Hare, M. Tomasello, "Human-like Social Skills in Dogs?" *Trends in Cognitive Sciences* 9, 439–44 (2005).

20. J. Riedel, K. Schumann, J. Kaminski, J. Call, M. Tomasello, "The Early Ontogeny of Human-Dog Communication," *Animal Behaviour* 75, 1003–14 (2008).

21. M. Gácsi, E. Kara, B. Belényi, J. Topál, Á. Miklósi, "The Effect of Development and Individual Differences in Pointing Comprehension of Dogs," *Animal Cognition* 12, 471–79 (2009).

22. B. Hare, M. Brown, C. Williamson, M. Tomasello, "The Domestication of Social Cognition in Dogs," *Science* 298, 1634–36 (2002).

23. J. Kaminski, L. Schulz, M. Tomasello, "How Dogs Know When Communication Is Intended for Them," *Developmental Science* 15, 222–32 (2012).

24. F. Rossano, M. Nitzschner, M. Tomasello, "Domestic Dogs and Puppies Can Use Human Voice Direction Referentially," *Proceedings of the Royal Society of London B: Biological Sciences* 281 (2014).

25. B. Hare, M. Tomasello, "Domestic Dogs (*Canis familiaris*) Use Human and Conspecific Social Cues to Locate Hidden Food," *Journal of Comparative Psychology* 113, 173 (1999).

26. G. Werhahn, Z. Virányi, G. Barrera, A. Sommese, F. Range, "Wolves (*Canis lupus*) and Dogs (*Canis familiaris*) Differ in Following Human Gaze into Distant Space but Respond Similarly to Their Packmates' Gaze," *Journal of Comparative Psychology* 130, 288 (2016).

27. F. Range, Z. Virányi, "Tracking the Evolutionary Origins of Dog-human Cooperation: The 'Canine Cooperation Hypothesis,'" *Frontiers in Psychology* 5, 1582 (2015).

28. Other researchers did even more extensive tests (Kaminski and Marshall-Pescini, 2014; Lampe, Bräuer, Kaminski, and Virányi, 2017; Marshall-Pescini, Rao, Virányi, and Range, 2017; Udell, Spencer, Dorey, and Wynne, 2012). The biologist Adam Miklosi compared a group of wolves and dogs that he raised in the same way. He immediately noticed several differences. Even as puppies, dogs used more communicative signals than wolves to engage and interact with their caretakers. Dog puppies whined, wagged their tails, and made eye contact with their caretakers and other people, while wolf puppies were anxious or even aggressive toward people, even their caretakers (Gácsi et al., 2009). Dog puppies made more eye contact with people than wolves did (Bentosela, Wynne, D'Orazio, Elgier, and Udell, 2016; Gácsi et al., 2009). In fact, when the dogs and wolves were given a container with food that they were unable to open, the dogs looked back toward their caretakers as if requesting help, while the wolves just kept trying to solve the problem on their own (Miklósi et al., 2003; Topál, Gergely, Erdőhegyi, Csibra, and Miklósi, 2009). When Adam tested the ability of wolves to follow a pointing gesture, they ignored the extended arm of their caretaker even though the same person had pointed things out to them countless times before.

Even with extensive training, the wolves were only at the same level as dogs, who were untrained at using human gestures. Our own research group has recently compared more than two dozen wolf puppies to two dozen age-matched dog puppies and again found that dogs outperform wolves when cooperatively communicating with humans but not in nonsocial tasks. All of this research led to one plausible explanation for how dogs came to be our best friends and champions of human-like cooperative communication. M. Bentosela, C.D.L. Wynne, M. D'Orazio, A. Elgier, M.A.R. Udell, "Sociability and Gazing Toward Humans in Dogs and Wolves: Simple Behaviors with Broad Implications," *Journal of the Experimental Analysis of Behavior* 105(1), 68–75 (2016); M. Gácsi, B. Gyoöri, Z. Virányi, E. Kubinyi, F. Range, B. Belényi, Á. Miklósi, "Explaining Dog Wolf Differences in Utilizing Human Pointing Gestures: Selection for Synergistic Shifts in the Development of Some Social Skills," *PLoS One* 4(8), e6584 (2009); Juliane Kaminski, Sarah Marshall-Pescini, *The Social Dog: Behavior and Cognition* (Elsevier, 2014); M. Lampe, J. Bräuer, J. Kaminski, Z. Virányi, "The Effects of Domestication and Ontogeny on Cognition in Dogs and Wolves," *Scientific Reports* 7(1), 11690 (2017); S. Marshall-Pescini, A. Rao, Z. Virányi, F. Range, "The Role of Domestication and Experience in 'Looking Back' Towards Humans in an Unsolvable Task," *Scientific Reports* 7 (2017); Á. Miklósi, E. Kubinyi, J. Topál, M. Gácsi, Z. Virányi, V. Csányi, "A Simple Reason for a Big Difference: Wolves Do Not Look Back at Humans, but Dogs Do," *Current Biology* 13(9), 763–66 (2003), 10:1016/S0960-9822(03)00263-X; J. Topál, G. Gergely, Á.Erdőhegyi, G. Csibra, Á. Miklósi, "Differential Sensitivity to Human Communication in Dogs, Wolves, and Human Infants," *Science* 325 (5945), 1269–72 (2009); M.A.R. Udell, J. M. Spencer, N. R. Dorey, C.D.L. Wynne, "Human-Socialized Wolves Follow Diverse Human Gestures . . . and They May Not Be Alone," *International Journal of Comparative Psychology* 25(2) (2012).
29. M.A.R. Udell, J. M. Spencer, N. R. Dorey, C.D.L. Wynne, "Human-Socialized Wolves Follow Diverse Human Gestures . . . and They May Not Be Alone," *International Journal of Comparative Psychology* 25(2) (2012).
30. M. Lampe, J. Bräuer, J. Kaminski, Z. Virányi, "The Effects of Domestication and Ontogeny on Cognition in Dogs and Wolves," *Scientific Reports* 7, 11690 (2017).

31. S. Marshall-Pescini, A. Rao, Z. Virányi, F. Range, "The Role of Domestication and Experience in 'Looking Back' Towards Humans in an Unsolvable Task," *Scientific Reports* 7 (2017).

32. Juliane Kaminski, Sarah Marshall-Pescini, *The Social Dog: Behavior and Cognition* (Elsevier, 2014).

33. S. Marshall-Pescini, J. Kaminski, "The Social Dog: History and Evolution," in *The Social Dog: Behavior and Cognition* (Elsevier, 2014), 3–33.

34. Google Trends (2015).

35. J. Butler, W. Brown, J. du Toit, "Anthropogenic Food Subsidy to a Commensal Carnivore: The Value and Supply of Human Faeces in the Diet of Free-Ranging Dogs," *Animals* 8, 67 (2018).

36. Steven E. Churchill, *Thin on the Ground: Neandertal Biology, Archeology, and Ecology* (Hoboken, NJ: John Wiley & Sons, 2014), vol. 10.

37. Pat Shipman, *The Invaders* (Cambridge, MA: Harvard University Press, 2015).

38. Raymond Coppinger, Lorna Coppinger, *Dogs: A New Understanding of Canine Origin, Behavior and Evolution* (Chicago: University of Chicago Press, 2002).

39. J. R. Butler, W. Y. Brown, J. T. du Toit, "Anthropogenic Food Subsidy to a Commensal Carnivore: The Value and Supply of Human Faeces in the Diet of Free-Ranging Dogs," *Animals* 8 (2018).

40. K. D. Lupo, "When and Where Do Dogs Improve Hunting Productivity? The Empirical Record and Some Implications for Early Upper Paleolithic Prey Acquisition," *Journal of Anthropological Archaeology* 47, 139–51 (2017).

41. B. P. Smith, C. A. Litchfield, "A Review of the Relationship Between Indigenous Australians, Dingoes (*Canis dingo*) and Domestic Dogs (*Canis familiaris*)," *Anthrozoös* 22, 111–28 (2009).

42. Stanley D. Gehrt, Seth P. D. Riley, and Brian L. Cypher, eds., *Urban Carnivores: Ecology, Conflict, and Conservation* (Baltimore: Johns Hopkins University Press, 2010) 79–95.

43. We worked in collaboration with Roland Kays of the Raleigh Museum of Natural Science, who runs the Candid Critters project. He organized citizen scientists to set up wildlife cameras all over the state to monitor wildlife behavior. The paper is currently under review.

44. E. L. MacLean, B. Hare, C. L. Nunn, E. Addessi, F. Amici, R. C.

Anderson, F. Aureli, J. M. Baker, A. E. Bania, A. M. Barnard, "The Evolution of Self-control," *Proceedings of the National Academy of Sciences* 111, E2140–E2148 (2014).

45. Gehrt et al., eds., *Urban Carnivores.*

46. J. Partecke, E. Gwinner, S. Bensch, "Is Urbanisation of European Blackbirds (*Turdus merula*) Associated with Genetic Differentiation?" *Journal of Ornithology* 147, 549–52 (2006).

47. J. Partecke, I. Schwabl, E. Gwinner, "Stress and the City: Urbanization and Its Effects on the Stress Physiology in European Blackbirds," *Ecology* 87, 1945–52 (2006).

48. P. M. Harveson, R. R. Lopez, B. A. Collier, N. J. Silvy, "Impacts of Urbanization on Florida Key Deer Behavior and Population Dynamics," *Biological Conservation* 134, 321–31 (2007).

49. R. McCoy, S. Murphie, "Factors Affecting the Survival of Black-tailed Deer Fawns on the Northwestern Olympic Peninsula, Washington," *Makah Tribal Forestry Final Report,* Neah Bay, Washington (2011).

50. A. Hernádi, A. Kis, B. Turcsán, J. Topál, "Man's Underground Best Friend: Domestic Ferrets, Unlike the Wild Forms, Show Evidence of Dog-like Social-Cognitive Skills," *PLoS One* 7, e43267 (2012).

51. K. Okanoya, "Sexual Communication and Domestication May Give Rise to the Signal Complexity Necessary for the Emergence of Language: An Indication from Songbird Studies," *Psychonomic Bulletin & Review* 24, 106–10 (2017).

52. R.T.T. Forman, "The Urban Region: Natural Systems in Our Place, Our Nourishment, Our Home Range, Our Future," *Landscape Ecology* 23, 251–53 (2008).

3 Our Long-Lost Cousins

1. B. Hare, V. Wobber, R. Wrangham, "The Self-Domestication Hypothesis: Evolution of Bonobo Psychology Is Due to Selection Against Aggression," *Animal Behaviour* 83, 573–85 (2012).

2. R. Wrangham, D. Pilbeam, "Apes as Time Machines," in *All Apes Great and Small* (New York: Springer, 2002), 5–17.

3. Richard W. Wrangham, Dale Peterson, *Demonic Males: Apes and the Origins of Human Violence* (Houghton Mifflin Harcourt, 1996).

4. M. L. Wilson, M. D. Hauser, R. W. Wrangham, "Does Partici-

pation in Intergroup Conflict Depend on Numerical Assessment, Range Location, or Rank for Wild Chimpanzees?" *Animal Behaviour* 61, 1203–16 (2001).

5. M. L. Wilson, C. Boesch, B. Fruth, T. Furuichi, I. C. Gilby, C. Hashimoto, C. L. Hobaiter, G. Hohmann, N. Itoh, K. Koops, "Lethal Aggression in Pan Is Better Explained by Adaptive Strategies than Human Impacts," *Nature* 513, 414–17 (2014).

6. J. C. Mitani, D. P. Watts, S. J. Amsler, "Lethal Intergroup Aggression Leads to Territorial Expansion in Wild Chimpanzees," *Current Biology* 20, R507–R8 (2010).

7. R. W. Wrangham, M. L. Wilson, "Collective Violence: Comparisons Between Youths and Chimpanzees," *Annals of the New York Academy of Sciences* 1036, 233–56 (2004).

8. S. M. Kahlenberg, M. E. Thompson, M. N. Muller, R. W. Wrangham, "Immigration Costs for Female Chimpanzees and Male Protection as an Immigrant Counterstrategy to Intrasexual Aggression," *Animal Behaviour* 76, 1497–1509 (2008).

9. Frans B. de Waal, F. Lanting, *Bonobo: The Forgotten Ape* (University of California Press, 1997).

10. K. Walker, B. Hare, in *Bonobos: Unique in Mind, Brain and Behavior,* edited by B. Hare and S. Yamamoto (Oxford University Press, 2017), chapter 4, 49–64.

11. Brian Hare, Shinya Yamamoto, *Bonobos: Unique in Mind, Brain, and Behavior* (Oxford University Press, 2017).

12. P. H. Douglas, G. Hohmann, R. Murtagh, R. Thiessen-Bock, T. Deschner, "Mixed Messages: Wild Female Bonobos Show High Variability in the Timing of Ovulation in Relation to Sexual Swelling Patterns," *BMC Evolutionary Biology* 16, 140 (2016).

13. T. Furuichi, "Female Contributions to the Peaceful Nature of Bonobo Society," *Evolutionary Anthropology: Issues, News, and Reviews* 20, 131–42 (2011).

14. N. Tokuyama, T. Furuichi, "Do Friends Help Each Other? Patterns of Female Coalition Formation in Wild Bonobos at Wamba," *Animal Behaviour* 119, 27–35 (2016).

15. L. R. Moscovice, M. Surbeck, B. Fruth, G. Hohmann, A. V. Jaeggi, T. Deschner, "The Cooperative Sex: Sexual Interactions Among Female Bonobos Are Linked to Increases in Oxytocin, Proximity and Coalitions," *Hormones and Behavior* 116, 104581 (2019).

16. Richard Wrangham, *The Goodness Paradox: The Strange Relation-*

ship Between Virtue and Violence in Human Evolution (New York: Pantheon, 2019).

17. There is one suspected case. A male bonobo was attacked by a number of bonobos and sustained serious wounds, and he was never seen again. However, the outcome was not confirmed and it is possible the male survived. M. L. Wilson, C. Boesch, B. Fruth, T. Furuichi, I. C. Gilby, C. Hashimoto, C. L. Hobaiter, G. Hohmann, N. Itoh, K.J.N. Koops, "Lethal Aggression in Pan Is Better Explained by Adaptive Strategies than Human Impacts," *Nature* 513, 414 (2014).

18. T. Sakamaki, H. Ryu, K. Toda, N. Tokuyama, T. Furuichi, "Increased Frequency of Intergroup Encounters in Wild Bonobos (*Pan Paniscus*) Around the Yearly Peak in Fruit Abundance at Wamba," *International Journal of Primatology* 3, 685–704 (2018); Brian Hare, Shinya Yamamoto, *Bonobos: Unique in Mind, Brain, and Behavior* (Oxford University Press, 2017).

19. M. Surbeck, R. Mundry, G. Hohmann, "Mothers Matter! Maternal Support, Dominance Status and Mating Success in Male Bonobos (*Pan paniscus*)," *Proceedings of the Royal Society of London B: Biological Sciences* 278, 590–98 (2011).

20. M. Surbeck, T. Deschner, G. Schubert, A. Weltring, G. Hohmann, "Mate Competition, Testosterone and Intersexual Relationships in Bonobos, *Pan paniscus*," *Animal Behaviour* 83(3), 659–69 (2012).

21. M. Surbeck, K. E. Langergraber, B. Fruth, L. Vigilant, G. Hohmann, "Male Reproductive Skew Is Higher in Bonobos than Chimpanzees," *Current Biology* 27, R640–R641 (2017).

22. S. Ishizuka, Y. Kawamoto, T. Sakamaki, N. Tokuyama, K. Toda, H. Okamura, T. Furuichi, "Paternity and Kin Structure Among Neighbouring Groups in Wild Bonobos at Wamba," *Royal Society Open Science* 5, 171006 (2018).

23. C. B. Stanford, "The Social Behavior of Chimpanzees and Bonobos: Empirical Evidence and Shifting Assumptions," *Current Anthropology* 39, 399–420 (1998).

24. B. Hare, S. Kwetuenda, "Bonobos Voluntarily Share Their Own Food with Others," *Current Biology* 20, R230–R231 (2010).

25. J. Tan, B. Hare, "Bonobos Share with Strangers," *PLoS One* 8, e51922 (2013).

26. J. Tan, D. Ariely, B. Hare, "Bonobos Respond Prosocially Toward Members of Other Groups," *Scientific Reports* 7, 14733 (2017).

27. V. Wobber, B. Hare, J. Maboto, S. Lipson, R. Wrangham, P. T. Ellison, "Differential Changes in Steroid Hormones Before Competition in Bonobos and Chimpanzees," *Proceedings of the National Academy of Sciences* 107, 12457–62 (2010).

28. M. H. McIntyre, E. Herrmann, V. Wobber, M. Halbwax, C. Mohamba, N. de Sousa, R. Atencia, D. Cox, B. Hare, "Bonobos Have a More Human-like Second-to-Fourth Finger Length Ratio (2D:4D) than Chimpanzees: A Hypothesized Indication of Lower Prenatal Androgens," *Journal of Human Evolution* 56, 361–65 (2009).

29. C. D. Stimpson, N. Barger, J. P. Taglialatela, A. Gendron-Fitzpatrick, P. R. Hof, W. D. Hopkins, C. C. Sherwood, "Differential Serotonergic Innervation of the Amygdala in Bonobos and Chimpanzees," *Social Cognitive and Affective Neuroscience* 11, 413–22 (2015).

30. C. H. Lew, K. L. Hanson, K. M. Groeniger, D. Greiner, D. Cuevas, B. Hrvoj-Mihic, C. M. Schumann, K. Semendeferi, "Serotonergic Innervation of the Human Amygdala and Evolutionary Implications," *American Journal of Physical Anthropology* 170, 351–360 (2019).

31. Lyudmila Trut, "Early Canid Domestication: The Farm-Fox Experiment; Foxes Bred for Tamability in a 40-year Experiment Exhibit Remarkable Transformations That Suggest an Interplay Between Behavioral Genetics and Development," *American Scientist* 87, 160–69 (1999).

32. B. Agnvall, J. Bélteky, R. Katajamaa, P. Jensen, "Is Evolution of Domestication Driven by Tameness? A Selective Review with Focus on Chickens," *Applied Animal Behaviour Science* (2017).

33. E. Herrmann, B. Hare, J. Call, M. Tomasello, "Differences in the Cognitive Skills of Bonobos and Chimpanzees," *PLoS One* 5, e12438 (2010).

34. When researchers used computer eye tracking to compare how bonobos and chimpanzees respond to human faces, they found that chimpanzees mostly focused on a person's mouth while ignoring their eyes. Bonobos focused primarily on the eyes of others and more than chimpanzees; F. Kano, J. Call, "Great Apes Generate Goal-Based Action Predictions: An Eye-Tracking Study," *Psychological Science* 25, 1691–98 (2014).

35. The anthropologist Zanna Clay found that unlike most animal calls, bonobo peeps can mean different things, either positive or negative; Z. Clay, A. Jahmaira, and K. Zuberbühler, "Functional Flexibility in Wild Bonobo Vocal Behavior," *PeerJ* 3, e1124 (2015).

36. A. P. Melis, B. Hare, M. Tomasello, "Chimpanzees Recruit the Best Collaborators," *Science* 311, 1297–1300 (2006).

37. A. P. Melis, B. Hare, M. Tomasello, "Chimpanzees Coordinate in a Negotiation Game," *Evolution and Human Behavior* 30, 381–92 (2009).

38. A. P. Melis, B. Hare, M. Tomasello, "Engineering Cooperation in Chimpanzees: Tolerance Constraints on Cooperation," *Animal Behaviour* 72, 275–86 (2006); B. Hare, A. P. Melis, V. Woods, S. Hastings, R. Wrangham, "Tolerance Allows Bonobos to Outperform Chimpanzees on a Cooperative Task," *Current Biology* 17, 619–23 (2007).

39. V. Wobber, R. Wrangham, B. Hare, "Bonobos Exhibit Delayed Development of Social Behavior and Cognition Relative to Chimpanzees," *Current Biology* 20, 226–30 (2010).

40. B. Hare, A. P. Melis, V. Woods, S. Hastings, R. Wrangham, "Tolerance Allows Bonobos to Outperform Chimpanzees on a Cooperative Task," *Current Biology* 17, 619–23 (2007).

4   Domesticated Minds

1. Jerry Kagan, Nancy Snidman, *The Long Shadow of Temperament* (Cambridge, MA: Harvard University Press, 2004).

2. C. E. Schwartz, C. I. Wright, L. M. Shin, J. Kagan, S. L. Rauch, "Inhibited and Uninhibited Infants 'Grown Up': Adult Amygdalar Response to Novelty," *Science* 300, 1952–53 (2003).

3. H. M. Wellman, J. D. Lane, J. LaBounty, S. L. Olson, "Observant, Nonaggressive Temperament Predicts Theory-of-mind Development," *Developmental Science* 14, 319–26 (2011).

4. Y.-T. Matsuda, K. Okanoya, M. Myowa-Yamakoshi, "Shyness in Early Infancy: Approach-Avoidance Conflicts in Temperament and Hypersensitivity to Eyes During Initial Gazes to Faces," *PLoS One* 8, e65476 (2013).

5. J. D. Lane, H. M. Wellman, S. L. Olson, A. L. Miller, L. Wang, T. Tardif, "Relations Between Temperament and Theory of Mind Development in the United States and China: Biological and Behavioral Correlates of Preschoolers' False-Belief Understanding," *Developmental Psychology* 49, 825–36 (2013).

6. Ibid., 825.

7. E. Longobardi, P. Spataro, M. D'Alessandro, R. Cerutti, "Temperament Dimensions in Preschool Children: Links with Cognitive and

Affective Theory of Mind," *Early Education and Development* 28, 377–95 (2017).

8. J. LaBounty, L. Bosse, S. Savicki, J. King, S. Eisenstat, "Relationship Between Social Cognition and Temperament in Preschool-Aged Children," *Infant and Child Development* 26, e1981 (2017).

9. A. V. Utevsky, D. V. Smith, S. A. Huettel, "Precuneus Is a Functional Core of the Default-Mode Network," *Journal of Neuroscience* 34, 932–40 (2014), 10:1523/JNEUROSCI.4227-13:2014.

10. R. M. Carter, S. A. Huettel, "A Nexus Model of the Temporal-Parietal Junction," *Trends in Cognitive Sciences* 17, 328–36 (2013).

11. H. Gweon, D. Dodell-Feder, M. Bedny, R. Saxe, "Theory of Mind Performance in Children Correlates with Functional Specialization of a Brain Region for Thinking About Thoughts," *Child Development* 83, 1853–68 (2012).

12. R. Saxe, S. Carey, N. Kanwisher, "Understanding Other Minds: Linking Developmental Psychology and Functional Neuroimaging," *Annual Review of Psychology* 55, 87–124 (2004).

13. E. G. Bruneau, N. Jacoby, R. Saxe, "Empathic Control Through Coordinated Interaction of Amygdala, Theory of Mind and Extended Pain Matrix Brain Regions," *Neuroimage* 114, 105–19 (2015).

14. F. Beyer, T. F. Münte, C. Erdmann, U. M. Krämer, "Emotional Reactivity to Threat Modulates Activity in Mentalizing Network During Aggression," *Social Cognitive and Affective Neuroscience* 9, 1552–60 (2013).

15. B. Hare, "Survival of the Friendliest: *Homo sapiens* Evolved via Selection for Prosociality," *Annual Review of Psychology* 68, 155–86 (2017).

16. R. W. Wrangham, "Two Types of Aggression in Human Evolution," *Proceedings of the National Academy of Sciences* 201713611 (2017).

17. Richard Wrangham, *The Goodness Paradox: The Strange Relationship Between Virtue and Violence in Human Evolution* (New York: Pantheon, 2019).

18. T. A. Hare, C. F. Camerer, A. Rangel, "Self-control in Decision-making Involves Modulation of the vmPFC Valuation System," *Science* 324, 646–48 (2009).

19. W. Mischel, Y. Shoda, P. K. Peake, "The Nature of Adolescent Competencies Predicted by Preschool Delay of Gratification," *Journal of Personality and Social Psychology* 54, 687 (1988).

20. T. W. Watts, G. J. Duncan, H. Quan, "Revisiting the Marshmallow Test: A Conceptual Replication Investigating Links Between Early Delay of Gratification and Later Outcomes," *Psychological Science* 29, 1159–77 (2018).

21. L. Michaelson, Y. Munakata, "Same Dataset, Different Conclusions: Preschool Delay of Gratification Predicts Later Behavioral Outcomes in a Preregistered Study," *Psychological Science* (in press).

22. T. E. Moffitt, L. Arseneault, D. Belsky, N. Dickson, R. J. Hancox, H. Harrington, R. Houts, R. Poulton, B. W. Roberts, S. Ross, "A Gradient of Childhood Self-control Predicts Health, Wealth, and Public Safety," *Proceedings of the National Academy of Sciences* 108, 2693–98 (2011).

23. E. L. MacLean, B. Hare, C. L. Nunn, E. Addessi, F. Amici, R. C. Anderson, F. Aureli, J. M. Baker, A. E. Bania, A. M. Barnard, "The Evolution of Self-control," *Proceedings of the National Academy of Sciences* 111, E2140–48 (2014).

24. Suzanna Herculano-Houzel, *The Human Advantage: A New Understanding of How Our Brain Became Remarkable* (Cambridge, MA: MIT Press, 2016).

25. M. Grabowski, B. Costa, D. Rossoni, G. Marroig, J. DeSilva, S. Herculano-Houzel, S. Neubauer, M. Grabowski, "From Bigger Brains to Bigger Bodies: The Correlated Evolution of Human Brain and Body Size," *Current Anthropology* 57, (2016).

26. S. Herculano-Houzel, "The Remarkable, yet Not Extraordinary, Human Brain as a Scaled-up Primate Brain and Its Associated Cost," *Proceedings of the National Academy of Sciences* 109, 10661–68 (2012), 10:1073/pnas.1201895109.

27. R. Holloway, "The Evolution of the Hominid Brain" in *Handbook of Paleoanthropology,* edited by W. Henke, I. Tattersall (Springer-Verlag, 2015), 1961–87.

28. Michael Tomasello, *Becoming Human: A Theory of Ontogeny* (Cambridge, MA: Belknap Press of Harvard University Press, 2019).

29. Joseph Henrich, *The Secret of Our Success: How Culture Is Driving Human Evolution, Domesticating Our Species, and Making Us Smarter* (Princeton, NJ: Princeton University Press, 2015).

30. M. Muthukrishna, B. W. Shulman, V. Vasilescu, J. Henrich, "Sociality Influences Cultural Complexity," *Proceedings of the Royal Society of London B: Biological Sciences* 281, 20132511 (2014).

31. D. W. Bird, R. B. Bird, B. F. Codding, D. W. Zeanah, "Variability

in the Organization and Size of Hunter-gatherer Groups: Foragers Do Not Live in Small-Scale Societies," *Journal of Human Evolution* 131, 96–108 (2019).

32. K. R. Hill, B. M. Wood, J. Baggio, A. M. Hurtado, R. T. Boyd, "Hunter-gatherer Inter-band Interaction Rates: Implications for Cumulative Culture," *PLoS One* 9, e102806 (2014).

33. A. Powell, S. Shennan, M. G. Thomas, "Late Pleistocene Demography and the Appearance of Modern Human Behavior," *Science* 324, 1298–1301 (2009).

34. R. L. Cieri, S. E. Churchill, R. G. Franciscus, J. Tan, B. Hare, "Craniofacial Feminization, Social Tolerance, and the Origins of Behavioral Modernity," *Current Anthropology* 55, 419–43 (2014).

35. Richard Wrangham has argued for evidence of human self-domestication already being apparent much earlier, at the very origin of *Homo sapiens* at around 300,000 years ago. Richard Wrangham, *The Goodness Paradox: The Strange Relationship Between Virtue and Violence in Human Evolution* (New York: Pantheon, 2019). It may be that the process ramped up over evolutionary time as human population densities increased. The main challenge to the hypothesis as I have constructed it would be if we saw evidence of self-domestication only after the appearance of modern human behavior, around 50,000 to 25,000 years ago. Another related problem is presented by new genomic comparisons suggesting that different human populations had already separated as early as 280,000 to 300,000 years ago—although perhaps a level of gene flow remained. David Reich, *Who We Are and How We Got Here: Ancient DNA and the New Science of the Human Past* (New York: Pantheon, 2018).

36. S. W. Gangestad, R. Thornhill, "Facial Masculinity and Fluctuating Asymmetry," *Evolution and Human Behavior* 24, 231–41 (2003).

37. B. Fink, K. Grammer, P. Mitteroecker, P. Gunz, K. Schaefer, F. L. Bookstein, J. T. Manning, "Second to Fourth Digit Ratio and Face Shape," *Proceedings of the Royal Society of London B: Biological Sciences* 272, 1995–2001 (2005).

38. J. C. Wingfield, "The Challenge Hypothesis: Where It Began and Relevance to Humans," *Hormones and Behavior* 92, 9–12 (2016).

39. P. B. Gray, J. F. Chapman, T. C. Burnham, M. H. McIntyre, S. F. Lipson, P. T. Ellison, "Human Male Pair Bonding and Testosterone," *Human Nature* 15, 119–31 (2004).

40. G. Rhodes, G. Morley, L. W. Simmons, "Women Can Judge

Sexual Unfaithfulness from Unfamiliar Men's Faces," *Biology Letters* 9, 20120908 (2013).

41. L. M. DeBruine, B. C. Jones, J. R. Crawford, L. L. M. Welling, A. C. Little, "The Health of a Nation Predicts Their Mate Preferences: Cross-cultural Variation in Women's Preferences for Masculinized Male Faces," *Proceedings of the Royal Society of London B: Biological Sciences* 277, 2405–10 (2010).

42. A. Sell, L. Cosmides, J. Tooby, D. Sznycer, C. von Rueden, M. Gurven, "Human Adaptations for the Visual Assessment of Strength and Fighting Ability from the Body and Face," *Proceedings of the Royal Society of London B: Biological Sciences* 276, 575–84 (2009).

43. B.T. Gleeson, "Masculinity and the Mechanisms of Human Self-domestication," *bioRxiv* 143875 (2018).

44. B. T. Gleeson, G.J.A. Kushnick, "Female Status, Food Security, and Stature Sexual Dimorphism: Testing Mate Choice as a Mechanism in Human Self-domestication," *American Journal of Physical Anthropology* 167, 458–469 (2018).

45. E. Nelson, C. Rolian, L. Cashmore, S. Shultz, "Digit Ratios Predict Polygyny in Early Apes, Ardipithecus, Neanderthals and Early Modern Humans but Not in Australopithecus," *Proceedings of the Royal Society B* (2011), vol. 278, 1556–63.

46. D. Kruska, "Mammalian Domestication and its Effect on Brain Structure and Behavior" in *Intelligence and Evolutionary Biology* (New York: Springer, 1988), 211–50.

47. H. Leach, C. Groves, T. O'Connor, O. Pearson, M. Zeder, H. Leach, "Human Domestication Reconsidered," *Current Anthropology* 44, 349–68 (2003).

48. N. K. Popova, "From Genes to Aggressive Behavior: The Role of the Serotonergic System," *Bioessays* 28, 495–503 (2006).

49. H. V. Curran, H. Rees, T. Hoare, R. Hoshi, A. Bond, "Empathy and Aggression: Two Faces of Ecstasy? A Study of Interpretative Cognitive Bias and Mood Change in Ecstasy Users," *Psychopharmacology* 173, 425–33 (2004).

50. E. F. Coccaro, L. J. Siever, H. M. Klar, G. Maurer, K. Cochrane, T. B. Cooper, R. C. Mohs, K. L. Davis, "Serotonergic Studies in Patients with Affective and Personality Disorders: Correlates with Suicidal and Impulsive Aggressive Behavior," *Archives of General Psychiatry* 46, 587–99 (1989).

51. M. J. Crockett, L. Clark, M. D. Hauser, T. W. Robbins, "Serotonin Selectively Influences Moral Judgment and Behavior Through Effects on Harm Aversion," *Proceedings of the National Academy of Sciences* 107, 17433–38 (2010).

52. A. Brumm, F. Aziz, G. D. Van den Bergh, M. J. Morwood, M. W. Moore, I. Kurniawan, D. R. Hobbs, R. Fullagar, "Early Stone Technology on Flores and Its Implications for *Homo floresiensis*," *Nature* 441, 624–28 (2006).

53. S. Alwan, J. Reefhuis, S. A. Rasmussen, R. S. Olney, J. M. Friedman, "Use of Selective Serotonin-Reuptake Inhibitors in Pregnancy and the Risk of Birth Defects," *New England Journal of Medicine* 356, 2684–92 (2007).

54. J. J. Cray, S. M. Weinberg, T. E. Parsons, R. N. Howie, M. Elsalanty, J. C. Yu, "Selective Serotonin Reuptake Inhibitor Exposure Alters Osteoblast Gene Expression and Craniofacial Development in Mice," *Birth Defects Research Part A: Clinical and Molecular Teratology* 100, 912–23 (2014).

55. C. Vichier-Guerre, M. Parker, Y. Pomerantz, R. H. Finnell, R. M. Cabrera, "Impact of Selective Serotonin Reuptake Inhibitors on Neural Crest Stem Cell Formation," *Toxicology Letters* 281, 20–25 (2017).

56. S. Neubauer, J. J. Hublin, P. Gunz, "The Evolution of Modern Human Brain Shape," *Science Advances* 4, eaao5961 (2018), published online EpubJan, 10:1126/sciadv.aao5961.

57. P. Gunz, A. K. Tilot, K. Wittfeld, A. Teumer, C. Y. Shapland, T. G. Van Erp, M. Dannemann, B. Vernot, S. Neubauer, T. Guadalupe, "Neandertal Introgression Sheds Light on Modern Human Endocranial Globularity," *Current Biology* 29, 120–27. e125 (2019).

58. A. Benítez-Burraco, C. Theofanopoulou, C. Boeckx, "Globularization and Domestication," *Topoi* 37, 265–278 (2016).

59. J.-J. Hublin, S. Neubauer, P. Gunz, "Brain Ontogeny and Life History in Pleistocene hominins," *Philosophical Transactions of the Royal Society B* 370, 20140062 (2015).

60. J. J. Negro, M. C. Blázquez, I. Galván, "Intraspecific Eye Color Variability in Birds and Mammals: A Recent Evolutionary Event Exclusive to Humans and Domestic Animals," *Frontiers in Zoology* 14, 53 (2017).

61. H. Kobayashi, S. Kohshima, "Unique Morphology of the Human Eye," *Nature* 387, 767 (1997).

62. T. Farroni, G. Csibra, F. Simion, M. H. Johnson, "Eye Contact

Detection in Humans From Birth," *Proceedings of the National Academy of Sciences* 99, 9602–9605 (2002).

63. E. L. MacLean, B. Hare, "Dogs Hijack the Human Bonding Pathway," *Science* 348, 280–81 (2015).

64. T. Farroni, S. Massaccesi, D. Pividori, M. H. Johnson, "Gaze Following in Newborns," *Infancy* 5, 39–60 (2004).

65. M. Carpenter, K. Nagell, M. Tomasello, G. Butterworth, C. Moore, "Social Cognition, Joint Attention, and Communicative Competence from 9 to 15 Months of Age," *Monographs of the Society for Research in Child Development* 63, i-174 (1998), 10:2307/1166214.

66. Michael Tomasello, *Constructing a Language* (Cambridge, MA: Harvard University Press, 2009).

67. N. L. Segal, A. T. Goetz, A. C. Maldonado, "Preferences for Visible White Sclera in Adults, Children, and Autism Spectrum Disorder Children: Implications of the Cooperative Eye Hypothesis," *Evolution and Human Behavior* 37, 35–39 (2016).

68. M. Tomasello, B. Hare, H. Lehmann, J. Call, "Reliance on Head Versus Eyes in the Gaze Following of Great Apes and Human Infants: The Cooperative Eye Hypothesis," *Journal of Human Evolution* 52, 314–20 (2007).

69. T. Grossmann, M. H. Johnson, S. Lloyd-Fox, A. Blasi, F. Deligianni, C. Elwell, G. Csibra, "Early Cortical Specialization for Face-to-face Communication in Human Infants," *Proceedings of the Royal Society of London B: Biological Sciences* 275, 2803–11 (2008).

70. T. C. Burnham, B. Hare, "Engineering Human Cooperation," *Human Nature* 18, 88–108 (2007).

71. P. J. Whalen, J. Kagan, R. G. Cook, F. C. Davis, H. Kim, S. Polis, D. G. McLaren, L. H. Somerville, A. A. McLean, J. S. Maxwell, "Human Amygdala Responsivity to Masked Fearful Eye Whites," *Science* 306, 2061–61 (2004).

72. Some work has raised questions about the strength of this link; S. B. Northover, W. C. Pedersen, A. B. Cohen, P. W. Andrews, "Artificial Surveillance Cues Do Not Increase Generosity: Two Meta-analyses," *Evolution and Human Behavior* 38, 144–53 (2017).

73. On balance the evidence seems to be in favor of eye contact promoting cooperation; C. Kelsey, A. Vaish, T.J.H.N. Grossmann, "Eyes, More than Other Facial Features, Enhance Real-World Donation Behavior," *Human Nature* 29, 390–401 (2018).

74. S. J. Gould, "A Biological Homage to Mickey Mouse," *Ecotone* 4, 333–40 (2008).

5 Forever Young

1. Stephen Jay Gould, *Ontogeny and Phylogeny* (Cambridge, MA: Harvard University Press, 1977).

2. Mary Jane West-Eberhard, *Developmental Plasticity and Evolution* (Oxford University Press, 2003).

3. C. A. Nalepa, C. Bandi, "Characterizing the Ancestors: Peadomorphosis and Termite Evolution," in *Termites: Evolution, Sociality, Symbioses, Ecology* (New York: Springer, 2000), 53–75.

4. M. F. Lawton, R. O. Lawton, "Heterochrony, Deferred Breeding, and Avian Sociality," in *Current Ornithology* 3 (New York: Plenum Press, 1986), 187–222.

5. J.-L. Gariépy, D. J. Bauer, R. B. Cairns, "Selective Breeding for Differential Aggression in Mice Provides Evidence for Heterochrony in Social Behaviours," *Animal Behaviour* 61, 933–47 (2001).

6. K. L. Cheney, R. Bshary, A.S.J.B.E. Grutter, "Cleaner Fish Cause Predators to Reduce Aggression Toward Bystanders at Cleaning Stations," *Behavioral Ecology* 19, 1063–67 (2008).

7. V. B. Baliga, R.S. Mehta, "Phylo-Allometric Analyses Showcase the Interplay Between Life-History Patterns and Phenotypic Convergence in Cleaner Wrasses," *The American Naturalist* 191, E129–43 (2018).

8. S. Gingins, R. Bshary, "The Cleaner Wrasse Outperforms Other Labrids in Ecologically Relevant Contexts, but Not in Spatial Discrimination," *Animal Behaviour* 115, 145–55 (2016).

9. A. Pinto, J. Oates, A. Grutter, R. Bshary, "Cleaner Wrasses *Labroides dimidiatus* Are More Cooperative in the Presence of an Audience," *Current Biology* 21, 1140–44 (2011).

10. Z. Triki, R. Bshary, A. S. Grutter, A. F. Ros, "The Argininevasotocin and Serotonergic Systems Affect Interspecific Social Behaviour of Client Fish in Marine Cleaning Mutualism," *Physiology & Behavior* 174, 136–43 (2017).

11. J. R. Paula, J. P. Messias, A. S. Grutter, R. Bshary, M.C.J.B.E. Soares, "The Role of Serotonin in the Modulation of Cooperative Behavior," *Behavioral Ecology* 26, 1005–12 (2015).

12. M. Gácsi, B. Győri, Á. Miklósi, Z. Virányi, E. Kubinyi, J. Topál, V. Csányi, "Species-specific Differences and Similarities in the Behavior of Hand-raised Dog and Wolf Pups in Social Situations with Humans," *Developmental Psychobiology: The Journal of the International Society for Developmental Psychobiology* 47, 111–22 (2005).

13. J. P. Scott, "The Process of Primary Socialization in Canine and Human Infants," *Monographs of the Society for Research in Child Development*, 1–47 (1963).

14. C. Hansen Wheat, W. van der Bijl, H. Temrin, "Dogs, but Not Wolves, Lose Their Sensitivity Toward Novelty with Age," *Frontiers in Psychology* 10, e2001–e2001 (2019).

15. Brian Hare, Vanessa Woods, *The Genius of Dogs* (Oneworld Publications, 2013).

16. D. Belyaev, I. Plyusnina, L. Trut, "Domestication in the Silver Fox (Vulpes fulvus Desm): Changes in Physiological Boundaries of the Sensitive Period of Primary Socialization," *Applied Animal Behaviour Science* 13, 359–70 (1985).

17. Lyudmila Trut, "Early Canid Domestication: The Farm-Fox Experiment; Foxes Bred for Tamability in a 40-year Experiment Exhibit Remarkable Transformations That Suggest an Interplay Between Behavioral Genetics and Development," *American Scientist* 87, 160–69 (1999).

18. Vanessa Woods, Brian Hare, "Bonobo but Not Chimpanzee Infants Use Socio-Sexual Contact with Peers," *Primates* 52, 111–16 (2011).

19. V. Wobber, B. Hare, S. Lipson, R. Wrangham, P. Ellison, "Different Ontogenetic Patterns of Testosterone Production Reflect Divergent Male Reproductive Strategies in Chimpanzees and Bonobos," *Physiology and Behavior,* 116, 44–53 (2013).

20. A similar pattern has been observed with the bonobo thyroid hormone. V. Behringer, T. Deschner, R. Murtagh, J. M. Stevens, G. Hohmann, "Age-related Changes in Thyroid Hormone Levels of Bonobos and Chimpanzees Indicate Heterochrony in Development," *Journal of Human Evolution* 66, 83–88 (2014).

21. Brian Hare, Shinya Yamamoto, *Bonobos: Unique in Mind, Brain, and Behavior* (Oxford University Press, 2017).

22. Just as our model predicts, a number of fish species are already prime candidates for self-domestication since changes in their developmental library genes cause friendliness and seemingly unrelated by-products. Take the species of blind cavefish known as Astyanax. This fish evolved from a river-dwelling species. Cavefish evolved in pools that lack both light and predators. Experiments have shown that the river species who must defend against predation are ten times more aggressive than cavefish. Meanwhile, cavefish have more sensitive noses and taste buds that make them four times more effi-

cient at foraging in the dark than their sighted relatives. Remarkably, all of these differences have been traced to a librarian gene (called sonic hedgehog) that changes how serotonin is made available in the embryonic brains of cavefish. As cavefish grow, they produce more serotonin and are more receptive to it in parts of their brain responsible for their lack of aggression and enhanced sense of smell and taste. The same process is responsible for their underdeveloped eyes that leave them blind. Selection for friendliness favored fish with brains that developed with more serotonin earlier. This early developmental change then drove both the behavioral and morphological changes that make these less aggressive blind fish so successful. All these changes are linked to a librarian gene in control of development. S. Rétaux, Y. Elipot, "Feed or Fight: A Behavioral Shift in Blind Cavefish," *Communicative & Integrative Biology* 6(2) (2013), 1–10.

23. A. S. Wilkins, R. W. Wrangham, W. T. Fitch, "The 'Domestication Syndrome' in Mammals: A Unified Explanation Based on Neural Crest Cell Behavior and Genetics," *Genetics* 197, 795–808 (2014).
24. G. W. Calloni, N. M. Le Douarin, E. Dupin, "High Frequency of Cephalic Neural Crest Cells Shows Coexistence of Neurogenic, Melanogenic, and Osteogenic Differentiation Capacities," *Proceedings of the National Academy of Sciences* 106, 8947–52 (2009).
25. C. Vichier-Guerre, M. Parker, Y. Pomerantz, R. H. Finnell, R. M. Cabrera, "Impact of Selective Serotonin Reuptake Inhibitors on Neural Crest Stem Cell Formation," *Toxicology Letters* 281, 20–25 (2017).
26. In support of this idea, comparisons between wolves and village dogs (i.e., dogs that have not experienced intense intentional selection for their appearance or behavior) reveal that neural crest genes have been under selection during the process of domestication. Future research will likely find that animals from wrasse to bonobos have been altered similarly. A. R. Boyko, R. H. Boyko, C. M. Boyko, H. G. Parker, M. Castelhano, L. Corey, . . . R. J. Kityo, "Complex Population Structure in African Village Dogs and Its Implications for Inferring Dog Domestication History," *Proceedings of the National Academy of Sciences* 0902129106 (2009).
27. C. Theofanopoulou, S. Gastaldon, T. O'Rourke, B. D. Samuels, A. Messner, P. T. Martins, F. Delogu, S. Alamri, C. Boeckx, "Self-domestication in *Homo sapiens*: Insights from Comparative Genomics," *PLoS One* 12, e0185306 (2017).

28. M. Zanella, A. Vitriolo, A. Andirko, P. T. Martins, S. Sturm, T. O'Rourke, M. Laugsch, N. Malerba, A. Skaros, S. Trattaro, "Dosage Analysis of the 7q11. 23 Williams Region identifies BAZ1B as a Master Regulator of the Modern Human Face and Validate the Self-Domestication Hypothesis," *Science Advances* 5, 12 (2019).

29. Brian Hare, "Survival of the Friendliest: *Homo sapiens* Evolved via Selection for Prosociality," *Annual Review of Psychology* 68, 155–86 (2017).

30. Martin N. Muller, Richard Wrangham, David Pilbeam, *Chimpanzees and Human Evolution* (Cambridge, MA: Harvard University Press, 2017).

31. J.-J. Hublin, S. Neubauer, P. Gunz, "Brain Ontogeny and Life History in Pleistocene hominins," *Philosophical Transactions of the Royal Society B: Biological Sciences* 370, 20140062 (2015).

32. V. Wobber, E. Herrmann, B. Hare, R. Wrangham, M. Tomasello, "Differences in the Early Cognitive Development of Children and Great Apes," *Developmental Psychobiology* 56, 547–73 (2014).

33. P. Gunz, S. Neubauer, L. Golovanova, V. Doronichev, B. Maureille, J.-J. Hublin, "A Uniquely Modern Human Pattern of Endocranial Development: Insights from a New Cranial Reconstruction of the Neandertal Newborn from Mezmaiskaya," *Journal of Human Evolution* 62, 300–13 (2012).

34. C. W. Kuzawa, H. T. Chugani, L. I. Grossman, L. Lipovich, O. Muzik, P. R. Hof, D. E. Wildman, C. C. Sherwood, W. R. Leonard, N. Lange, "Metabolic Costs and Evolutionary Implications of Human Brain Development," *Proceedings of the National Academy of Sciences* 111, 13010–15 (2014).

35. E. Bruner, T. M. Preuss, X. Chen, J. K. Rilling, "Evidence for Expansion of the Precuneus in Human Evolution," *Brain Structure and Function* 222, 1053–60 (2017).

36. T. Grossmann, M. H. Johnson, S. Lloyd-Fox, A. Blasi, F. Deligianni, C. Elwell, G. Csibra, "Early Cortical Specialization for Face-to-face Communication in Human Infants," *Proceedings of the Royal Society of London B: Biological Sciences* 275, 2803–11 (2008).

37. P. H. Vlamings, B. Hare, J. Call, "Reaching Around Barriers: The Performance of the Great Apes and 3–5-Year-Old Children," *Animal Cognition* 13, 273–85 (2010).

38. E. Herrmann, A. Misch, V. Hernandez-Lloreda, M. Tomasello, "Uniquely Human Self-control Begins at School Age," *Developmental Science* 18, 979–93 (2015).

39. B. Casey, "Beyond Simple Models of Self-control to Circuit-Based Accounts of Adolescent Behavior," *Annual Review of Psychology* 66, 295–319 (2015).

40. R. B. Bird, D. W. Bird, B. F. Codding, C. H. Parker, J. H. Jones, "The 'Fire Stick Farming' Hypothesis: Australian Aboriginal Foraging Strategies, Biodiversity, and Anthropogenic Fire Mosaics," *Proceedings of the National Academy of Sciences* 105, 14796–801 (2008).

41. J. C. Berbesque, B. M. Wood, A. N. Crittenden, A. Mabulla, F. W. Marlowe, "Eat First, Share Later: Hadza Hunter-gatherer Men Consume More While Foraging than in Central Places," *Evolution and Human Behavior* 37, 281–86 (2016).

42. M. Gurven, W. Allen-Arave, K. Hill, M. Hurtado, " 'It's a Wonderful Life': Signaling Generosity Among the Ache of Paraguay," *Evolution and Human Behavior* 21, 263–82 (2000).

43. C. Boehm, H. B. Barclay, R. K. Dentan, M.-C. Dupre, J. D. Hill, S. Kent, B. M. Knauft, K. F. Otterbein, S. Rayner, "Egalitarian Behavior and Reverse Dominance Hierarchy" [and comments and reply], *Current Anthropology* 34, 227–54 (1993).

44. M. J. Platow, M. Foddy, T. Yamagishi, L. Lim, A. Chow, "Two Experimental Tests of Trust in In-group Strangers: The Moderating Role of Common Knowledge of Group Membership," *European Journal of Social Psychology* 42, 30–35 (2012).

45. A. C. Pisor, M. Gurven, "Risk Buffering and Resource Access Shape Valuation of Out-group Strangers," *Scientific Reports* 6, 30435 (2016).

46. A. Romano, D. Balliet, T. Yamagishi, J. H. Liu, "Parochial Trust and Cooperation Across 17 Societies," *Proceedings of the National Academy of Sciences* 114, 12702–707 (2017).

47. J. K. Hamlin, N. Mahajan, Z. Liberman, K. Wynn, "Not Like Me = Bad: Infants Prefer Those Who Harm Dissimilar Others," *Psychological Science* 24, 589–94 (2013).

48. G. Soley, N. Sebastián-Gallés, "Infants Prefer Tunes Previously Introduced by Speakers of Their Native Language," *Child Development* 86, 1685–92 (2015).

49. N. McLoughlin, S. P. Tipper, H. Over, "Young Children Perceive Less Humanness in Outgroup Faces," *Developmental Science* 21, e12539 (2017).

50. L. M. Hackel, C. E. Looser, J. J. Van Bavel, "Group Membership Alters the Threshold for Mind Perception: The Role of Social

Identity, Collective Identification, and Intergroup Threat," *Journal of Experimental Social Psychology* 52, 15–23 (2014), published online Epub2014/05/01/, 10:1016/j.jesp.2013:12.001.

51. E. Sparks, M. G. Schinkel, C. Moore, "Affiliation Affects Generosity in Young Children: The Roles of Minimal Group Membership and Shared Interests," *Journal of Experimental Child Psychology* 159, 242–62 (2017).

52. J. S. McClung, S.D. Reicher, "Representing Other Minds: Mental State Reference Is Moderated by Group Membership," *Journal of Experimental Social Psychology* 76, 385–92 (2018).

53. Joseph Henrich, *The Secret of Our Success: How Culture Is Driving Human Evolution, Domesticating Our Species, and Making Us Smarter* (Princeton, NJ: Princeton University Press, 2015).

54. Serotonin neurons mediate the effects of oxytocin. Serotonin receptor activity generates a feedback loop in which serotonin increases with oxytocin. Testosterone prevents the binding of oxytocin, which in turn decreases serotonin. Brian Hare, "Survival of the Friendliest: *Homo sapiens* Evolved via Selection for Prosociality," *Annual Review of Psychology* 68, 155–86 (2017).

55. M. L. Boccia, P. Petrusz, K. Suzuki, L. Marson, C. A. Pedersen, "Immunohistochemical Localization of Oxytocin Receptors in Human Brain," *Neuroscience* 253, 155–64 (2013).

56. C. K. De Dreu, "Oxytocin Modulates Cooperation Within and Competition Between Groups: An Integrative Review and Research Agenda," *Hormones and Behavior* 61, 419–28 (2012).

57. M. Nagasawa, T. Kikusui, T. Onaka, M. Ohta, "Dog's Gaze at Its Owner Increases Owner's Urinary Oxytocin During Social Interaction," *Hormones and Behavior* 55, 434–41 (2009).

58. K. M. Brethel-Haurwitz, K. O'Connell, E. M. Cardinale, M. Stoianova, S. A. Stoycos, L. M. Lozier, J. W. VanMeter, A. A. Marsh, "Amygdala–midbrain Connectivity Indicates a Role for the Mammalian Parental Care System in Human Altruism," *Proceedings of the Royal Society B: Biological Sciences* 284, 20171731 (2017).

59. C. Theofanopoulou, A. Andirko, C. Boeckx, "Oxytocin and Vasopressin Receptor Variants as a Window onto the Evolution of Human Prosociality," bioRxiv, 460584 (2018).

60. K. R. Hill, B. M. Wood, J. Baggio, A. M. Hurtado, R. T. Boyd, "Hunter-gatherer Inter-band Interaction Rates: Implications for Cumulative Culture," *PLoS One* 9, e102806 (2014).

61. K. Hill, "Altruistic Cooperation During Foraging by the Ache, and the Evolved Human Predisposition to Cooperate," *Human Nature* 13, 105–28 (2002).

62. Steven Pinker, *The Better Angels of Our Nature: Why Violence Has Declined* (Penguin Books, 2012).

63. Y. N. Harari, *Homo Deus: A Brief History of Tomorrow* (Random House, 2016).

64. R. C. Oka, M. Kissel, M. Golitko, S. G. Sheridan, N. C. Kim, A. Fuentes, "Population Is the Main Driver of War Group Size and Conflict Casualties," *Proceedings of the National Academy of Sciences* 114, E11101–E11110 (2017).

6 Not Quite Human

1. "'Burundi: The Gatumba Massacre: War Crimes and Political Agendas,'" (Human Rights Watch, 2004).

2. Brian Hare, Shinya Yamamoto, *Bonobos: Unique in Mind, Brain, and Behavior* (Oxford University Press, 2017).

3. O. J. Bosch, S. A. Krömer, P. J. Brunton, I. D. Neumann, "Release of Oxytocin in the Hypothalamic Paraventricular Nucleus, but Not Central Amygdala or Lateral Septum in Lactating Residents and Virgin Intruders During Maternal Defence," *Neuroscience* 124, 439–48 (2004).

4. C. F. Ferris, K. B. Foote, H. M. Meltser, M. G. Plenby, K. L. Smith, T. R. Insel, "Oxytocin in the Amygdala Facilitates Maternal Aggression," *Annals of the New York Academy of Sciences* 652, 456–57 (1992).

5. In humans, there is still debate over whether oxytocin makes people more cooperative toward their group members but more aggressive toward outsiders or whether oxytocin does not directly cause aggression toward outsiders but the elevated level of empathy and cooperation within group members can provoke resentment from outsiders that may later escalate to aggression (C. K. De Dreu, "Oxytocin Modulates Cooperation Within and Competition Between Groups: An Integrative Review and Research Agenda," *Hormones and Behavior* 61, 419–28 [2012]).

6. D. A. Baribeau, E. Anagnostou, "Oxytocin and Vasopressin: Linking Pituitary Neuropeptides and Their Receptors to Social Neurocircuits," *Frontiers in Neuroscience* 9, (2015).

7. K. M. Brethel-Haurwitz, K. O'Connell, E. M. Cardinale, M.

Stoianova, S. A. Stoycos, L. M. Lozier, J. W. VanMeter, A. A. Marsh, "Amygdala–Midbrain Connectivity Indicates a Role for the Mammalian Parental Care System in Human Altruism," *Proceedings of the Royal Society B: Biological Sciences* 284, 20171731 (2017).

8. S. T. Fiske, L. T. Harris, A. J. Cuddy, "Why Ordinary People Torture Enemy Prisoners," *Science* 306, 1482–83 (2004).

9. L. W. Chang, A. R. Krosch, M. Cikara, "Effects of Intergroup Threat on Mind, Brain, and Behavior," *Current Opinion in Psychology* 11, 69–73 (2016).

10. M. Hewstone, M. Rubin, H. Willis, "Intergroup Bias," *Annual Review of Psychology* 53, 575–604 (2002).

11. G. Soley, N. Sebastián-Gallés, "Infants Prefer Tunes Previously Introduced by Speakers of Their Native Language," *Child Development* 86, 1685–92 (2015).

12. D. J. Kelly, P. C. Quinn, A. M. Slater, K. Lee, A. Gibson, M. Smith, L. Ge, O. Pascalis, "Three-Month-Olds, but Not Newborns, Prefer Own-Race Faces," *Developmental Science* 8, F31–F36 (2005).

13. J. K. Hamlin, N. Mahajan, Z. Liberman, K. Wynn, "Not Like Me = Bad: Infants Prefer Those Who Harm Dissimilar Others," *Psychological Science* 24, 589–94 (2013).

14. M. F. Schmidt, H. Rakoczy, M. Tomasello, "Young Children Enforce Social Norms Selectively Depending on the Violator's Group Affiliation," *Cognition* 124, 325–33 (2012).

15. J. J. Jordan, K. McAuliffe, F. Warneken, "Development of In-group Favoritism in Children's Third-party Punishment of Selfishness," *Proceedings of the National Academy of Sciences* 111, 12710–715 (2014).

16. E. L. Paluck, D. P. Green, "Prejudice Reduction: What Works? A Review and Assessment of Research and Practice," *Annual Review of Psychology* 60, 339–67 (2009).

17. A. Bandura, B. Underwood, M. E. Fromson, "Disinhibition of Aggression Through Diffusion of Responsibility and Dehumanization of Victims," *Journal of Research in Personality* 9, 253–69 (1975).

18. Brian Hare, "Survival of the Friendliest: *Homo sapiens* Evolved via Selection for Prosociality," *Annual Review of Psychology* 68, 155–86 (2017).

19. T. Baumgartner, L. Götte, R. Gügler, E. Fehr, "The Mentalizing Network Orchestrates the Impact of Parochial Altruism on Social Norm Enforcement," *Human Brain Mapping* 33, 1452–69 (2012).

20. E. G. Bruneau, N. Jacoby, R. Saxe, "Empathic Control Through Coordinated Interaction of Amygdala, Theory of Mind and Extended Pain Matrix Brain Regions," *Neuroimage* 114, 105–19 (2015); E. Bruneau, N. Jacoby, N. Kteily, R. Saxe, "Denying Humanity: The Distinct Neural Correlates of Blatant Dehumanization," *Journal of Experimental Psychology: General* 147, 1078–1093 (2018).

21. M. L. Boccia, P. Petrusz, K. Suzuki, L. Marson, C. A. Pedersen, "Immunohistochemical Localization of Oxytocin Receptors in Human Brain," *Neuroscience* 253, 155–64 (2013).

22. C. S. Sripada, K. L. Phan, I. Labuschagne, R. Welsh, P. J. Nathan, A. G. Wood, "Oxytocin Enhances Resting-state Connectivity Between Amygdala and Medial Frontal Cortex," *International Journal of Neuropsychopharmacology* 16, 255–60 (2012).

23. M. Cikara, E. Bruneau, J. Van Bavel, R. Saxe, "Their Pain Gives Us Pleasure: How Intergroup Dynamics Shape Empathic Failures and Counter-empathic Responses," *Journal of Experimental Social Psychology* 55, 110–25 (2014).

24. Lasana Harris, *Invisible Mind: Flexible Social Cognition and Dehumanization* (Cambridge, MA: MIT Press, 2017); L. Harris, S. Fiske, "Social Neuroscience Evidence for Dehumanised Perception," *European Review of Social Psychology,* 20, 192–231 (2009).

25. H. Zhang, J. Gross, C. De Dreu, Y. Ma, "Oxytocin Promotes Coordinated Out-group Attack During Intergroup Conflict in Humans," *eLife* 8, e40698 (2019); One explanation for this lack of empathy is that intranasal oxytocin makes people respond to the emotions of outsiders more like psychopaths and less like extreme altruists. Abigail Marsh found that psychopaths are less sensitive to fear in the faces of strangers and extreme altruists are more sensitive to this fear. Abigail A. Marsh, *The Fear Factor: How One Emotion Connects Altruists, Psychopaths, and Everyone in Between* (New York: Hachette Book Group, 2017). Other researchers found that giving oxytocin to people from one ethnic group makes them less likely to perceive expressions of fear or pain in the people from a different ethnic group. X. Xu, X. Zuo, X. Wang, S. Han, "Do You Feel My Pain? Racial Group Membership Modulates Empathic Neural Responses," *Journal of Neuroscience* 29, 8525–29 (2009); F. Sheng, Y. Liu, B. Zhou, W. Zhou, S. Han, "Oxytocin Modulates the Racial Bias in Neural Responses to Others' Suffering," *Biological Psychology* 92, 380–86 (2013).

26. C. K. De Dreu, L. L. Greer, M. J. Handgraaf, S. Shalvi, G. A. Van Kleef, M. Baas, F. S. Ten Velden, E. Van Dijk, S. W. Feith, "The

Neuropeptide Oxytocin Regulates Parochial Altruism in Intergroup Conflict Among Humans," *Science* 328, 1408–11 (2010).

27. C. K. De Dreu, M. E. Kret, "Oxytocin Conditions Intergroup Relations Through Upregulated In-group Empathy, Cooperation, Conformity, and Defense," *Biological Psychiatry* 79, 165–73 (2016).

28. C. K. De Dreu, L. L. Greer, G. A. Van Kleef, S. Shalvi, M. J. Handgraaf, "Oxytocin Promotes Human Ethnocentrism," *Proceedings of the National Academy of Sciences* 108, 1262–66 (2011).

29. X. Xu, X. Zuo, X. Wang, S. Han, "Do You Feel My Pain? Racial Group Membership Modulates Empathic Neural Responses," *Journal of Neuroscience* 29, 8525–29 (2009).

30. F. Sheng, Y. Liu, B. Zhou, W. Zhou, S. Han, "Oxytocin Modulates the Racial Bias in Neural Responses to Others' Suffering," *Biological Psychology* 92, 380–86 (2013).

31. J. Levy, A. Goldstein, M. Influs, S. Masalha, O. Zagoory-Sharon, R. Feldman, "Adolescents Growing Up Amidst Intractable Conflict Attenuate Brain Response to Pain of Outgroup," *Proceedings of the National Academy of Sciences* 113, 13696–701 (2016).

32. R. W. Wrangham, "Two Types of Aggression in Human Evolution," *Proceedings of the National Academy of Sciences*, 201713611 (2017).

33. R. C. Oka, M. Kissel, M. Golitko, S. G. Sheridan, N. C. Kim, A. Fuentes, "Population Is the Main Driver of War Group Size and Conflict Casualties," *Proceedings of the National Academy of Sciences* 114, E11101–E11110 (2017).

34. D. Crowe, *War Crimes, Genocide, and Justice: A Global History* (New York: Springer, 2014).

35. David. L. Smith, *Less than Human: Why We Demean, Enslave, and Exterminate Others* (New York: St. Martin's Press, 2011).

36. D. Barringer, *Raining on Evolution's Parade* (New York: F&W Publications, 2006).

37. N. Kteily, E. Bruneau, A. Waytz, S. Cotterill, "The Ascent of Man: Theoretical and Empirical Evidence for Blatant Dehumanization," *Journal of Personality and Social Psychology* 109, 901 (2015).

38. People were also asked how much they agreed with the statement "Muslims bombed Boston. We as a planet need to wipe them off this world. Every one of them." While the vast majority of people strongly disagreed, there was also a significant shift toward agreeing after the Boston Marathon bombing.

39. E. Bruneau, N. Kteily, "The Enemy as Animal: Symmetric Dehu-

manization During Asymmetric Warfare," *PLoS One* 12, e0181422 (2017).

40. N. S. Kteily, E. Bruneau, "Darker Demons of Our Nature: The Need to (Re)Focus Attention on Blatant Forms of Dehumanization," *Current Directions in Psychological Science* 26, 487–94 (2017).

41. E. Bruneau, N. Jacoby, N. Kteily, R. Saxe, "Denying Humanity: The Distinct Neural Correlates of Blatant Dehumanization," *Journal of Experimental Psychology: General,* 147, 1078–1093 (2018).

42. N. Kteily, G. Hodson, E. Bruneau, "They See Us as Less Than Human: Metadehumanization Predicts Intergroup Conflict Via Reciprocal Dehumanization," *Journal of Personality and Social Psychology* 110, 343 (2016).

43. Experiments have revealed that when we dehumanize someone our brain tends to view each facial feature as separate from each other as if they are not even part of a face. Viewing someone's face like an object makes it easier to inflict harm on them. K. M. Fincher, P. E. Tetlock, M. W. Morris, "Interfacing with Faces: Perceptual Humanization and Dehumanization," *Current Directions in Psychological Science* 26, 288–93 (2017).

44. "Deception on Capitol Hill," *New York Times,* January 15, 1992.

45. J. R. MacArthur, "Remember Nayirah, Witness for Kuwait?," *New York Times* op-ed, January 6, 1992.

7 The Uncanny Valley

1. G. M. Lueong, *The Forest People Without a Forest: Development Paradoxes, Belonging and Participation of the Baka in East Cameroon* (Berghahn Books, 2016).

2. BBC News. Pygmy artists housed in Congo zoo in *BBC News.* (2007). Published online 13 July 2007, http://news.bbc.co.uk/2/hi/africa/6898241.stm.

3. F. E. Hoxie, "Red Man's Burden," *Antioch Review* 37, 326–42 (1979).

4. T. Buquet, paper presented at the International Medieval Congress, Leeds, 2011.

5. C. Niekerk, "Man and Orangutan in Eighteenth-Century Thinking: Retracing the Early History of Dutch and German Anthropology," *Monatshefte* 96, 477–502 (2004).

6. Tetsuro Matsuzawa, Tatyana Humle, Yamakoshi Sugiyama, *The*

Chimpanzees of Bossou and Nimba (Springer Science & Business Media, 2011).

7. M. Mori, "The Uncanny Valley," *Energy* 7, 33–35 (1970).

8. J. van Wyhe, P. C. Kjærgaard, "Going the Whole Orang: Darwin, Wallace and the Natural History of Orangutans," *Studies in History and Philosophy of Science Part C: Studies in History and Philosophy of Biological and Biomedical Sciences* 51, 53–63 (2015), published online Epub2015/06/01/, 10:1016/j.shpsc.2015:02.006.

9. D. Livingstone Smith, I. Panaitui, "Aping the Human Essence," in *Simianization: Apes, Gender, Class, and Race,* edited by W. D. Hund, C. W. Mills, S. Sebastiani (LIT Verlag Münster, 2015), vol. 6.

10. Wulf D. Hund, Charles W. Mills, Silvia Sebastiani, eds., *Simianization: Apes, Gender, Class, and Race* (LIT Verlag Münster, 2015), vol. 6.

11. J. Hunt, *On the Negro's Place in Nature* (Trübner, for the Anthropological Society, 1863).

12. Thomas Jefferson, "Notes on Virginia," in *The Life and Selected Writings of Thomas Jefferson* 187, 275 (New York: Modern Library, 1944).

13. K. Kenny, "Race, Violence, and Anti-Irish Sentiment in the Nineteenth Century," in *Making the Irish American: History and Heritage of the Irish in the United States,* 364–78 (New York: New York University Press, 2006); D. L. Smith, *Less than Human: Why We Demean, Enslave, and Exterminate Others* (New York: St. Martin's Press, 2011).

14. S. Affeldt, "Exterminating the Brute," in Hund et al., *Simianization.*

15. C. J. Williams, *Freedom & Justice: Four Decades of the Civil Rights Struggle as Seen by a Black Photographer of the Deep South* (Macon, GA: Mercer University Press, 1995).

16. David L. Smith, *Less than Human: Why We Demean, Enslave, and Exterminate Others* (New York: St. Martin's Press, 2011).

17. L. S. Newman, R. Erber, *Understanding Genocide: The Social Psychology of the Holocaust* (Oxford University Press, 2002).

18. D. J. Goldhagen, M. Wohlgelernter, "Hitler's Willing Executioners," *Society* 34, 32–37 (1997).

19. Hannah Arendt, *Eichmann in Jerusalem* (Penguin, 1963).

20. R. J. Rummel, *Statistics of Democide: Genocide and Mass Murder Since 1900* (LIT Verlag Münster, 1998), vol. 2.

21. G. Clark, "The Human-Relations Society and the Ideological Society," *Japan Foundation Newsletter* (1978).

22. V. L. Hamilton, J. Sanders, S. J. McKearney, "Orientations Toward Authority in an Authoritarian State: Moscow in 1990," *Personality and Social Psychology Bulletin* 21, 356–65 (1995).

23. D. Johnson, "Red Army Troops Raped Even Russian Women as They Freed Them from Camps," *The Daily Telegraph,* January 25 (2002).

24. D. Roithmayr, *Reproducing Racism: How Everyday Choices Lock in White Advantage* (New York: New York University Press, 2014).

25. R. L. Fleegler, "Theodore G. Bilbo and the Decline of Public Racism, 1938–1947," *Journal of Mississippi History* 68, 1–27 (2006).

26. G. M. Fredrickson, *Racism: A Short History* (Princeton, NJ: Princeton University Press, 2015).

27. In postwar Europe, "direct and open expression of racial prejudice has declined" (N. Akrami, B. Ekehammar, T. Araya, "Classical and Modern Racial Prejudice: A Study of Attitudes Toward Immigrants in Sweden," *European Journal of Social Psychology* 30, 521–32 [2000]); in Germany, "prejudice and authoritarian attitudes appeared to be on the decline in postwar generations" (D. Horrocks, E. Kolinsky, *Turkish Culture in German Society Today* [Berghahn Books, 1996], vol. 1); in Russia, "Soviet domestic policy changes and the course of international events presented [German Russians] with a host of opportunities" (E. J. Schmaltz, S. D. Sinner, "'You Will Die Under Ruins and Snow': The Soviet Repression of Russian Germans as a Case Study of Successful Genocide," *Journal of Genocide Research* 4, 327–56 [2002]); in Sweden, "changes in sociopolitical climate in general since the Second World and in particular the tendency of people to present themselves as non-prejudiced and socially or politically may prevent the expression of blatant racial prejudice"; and in the UK, "levels of racial prejudice are falling and are likely to fall further" (R. Ford, "Is Racial Prejudice Declining in Britain?" *British Journal of Sociology* 59, 609–36 [2008]).

28. Ford, "Is Racial Prejudice Declining in Britain?"

29. L. Huddy, S. Feldman, "On Assessing the Political Effects of Racial Prejudice," *Annual Review of Political Science* 12, 423–47 (2009).

30. A. T. Thernstrom and S. Thernstrom, "Taking Race out of the Race," in *Los Angeles Times,* March 2 (2008).

31. D. Horrocks, E. Kolinsky, *Turkish Culture in German Society Today* (Berghahn Books, 1996), vol. 1.

32. M. Augoustinos, C. Ahrens, J. M. Innes, "Stereotypes and Prejudice: The Australian Experience," *British Journal of Social Psychology* 33, 125–41 (1994).

33. U.S. Census Bureau (2017).

34. R. C. Hetey, J. L. Eberhardt, "Racial Disparities in Incarceration Increase Acceptance of Punitive Policies," *Psychological Science* 25, 1949–54 (2014).

35. K. Welch, "Black Criminal Stereotypes and Racial Profiling," *Journal of Contemporary Criminal Justice* 23, 276–88 (2007).

36. V. Hutchings, "Race, Punishment, and Public Opinion," *Perspectives on Politics* 13, 757 (2015).

37. K. T. Ponds, "The Trauma of Racism: America's Original Sin," *Reclaiming Children and Youth* 22, 22 (2013).

38. M. Clair, J. Denis, "Sociology of Racism," *International Encyclopedia of the Social and Behavioral Sciences*, 2nd ed. (Oxford: Elsevier, 2015).

39. N. Akrami, B. Ekehammar, T. Araya, "Classical and Modern Racial Prejudice: A Study of Attitudes Toward Immigrants in Sweden," *European Journal of Social Psychology* 30, 521–32 (2000).

40. This new racism can also include negative stereotypes about black people, and feelings of insecurity that black people pose a threat to white people's position in the racial hierarchy. P. M. Sniderman, E. G. Carmines, "Reaching Beyond Race," *PS: Political Science & Politics* 30, 466–71 (1997); L. Bobo, V. L. Hutchings, "Perceptions of Racial Group Competition: Extending Blumer's Theory of Group Position to a Multiracial Social Context," *American Sociological Review* December 1, 951–72 (1996). "Many researchers dealing with racial prejudice agree that its expression has become more subtle in modern society." N. Akrami, B. Ekehammar, T. Araya, "Classical and Modern Racial Prejudice: A Study of Attitudes Toward Immigrants in Sweden," *European Journal of Social Psychology* 30, 521–32 (2000). Instead of genocide, this new prejudice is expressed in "denial of continued discrimination antagonism toward minority group demands, and resentment about special favors for minority groups" (ibid.). As an example of this new racism, the criminologist Kelly Welch points to black criminal stereotypes. K. Welch, "Black Criminal Stereotypes and Racial Profiling," *Journal of Contemporary*

Criminal Justice 23, 276–88 (2007). The *Los Angeles Times* reports that penalties for crack cocaine (mostly used by black people) are harsher than powder cocaine (typically used by white people) (J. Katz, *Los Angeles Times*, 2000). According to this model, new prejudice represents a new type of culture that expresses itself in very different ways than the other older extinct form.

41. A. McCarthy, "Our Dangerous Drift from Reason" in *National Review.* Published online September 24, 2016, https://www.national review.com/2016/09/police-shootings-black-white-media-narrative -population-difference/.

42. "The problem is black criminal behavior," reported *The Washington Times* in 2014, "which is one manifestation of a black pathology that ultimately stems from the breakdown of the black family" (J. Riley, "What the left wont tell you about black crime," *Washington Times,* July 21, 2014). Others counter that the real cause of the problem for black communities is "social and economic islolation" ("Criminal Justice Fact Sheet," NAACP, 2019, https://www.naacp .org/criminal-justice-fact-sheet/.)

43. Gordon W. Allport, *The Nature of Prejudice* (Basic Books, 1979).

44. S. E. Asch, "Studies of Independence and Conformity: I. A Minority of One Against a Unanimous Majority," *Psychological Monographs: General and Applied* 70, 1 (1956).

45. S. Milgram, "The Perils of Obedience," *Harper's* 12 (1973).

46. A. Bandura, B. Underwood, M. E. Fromson, "Disinhibition of Aggression Through Diffusion of Responsibility and Dehumanization of Victims," *Journal of Research in Personality* 9, 253–69 (1975).

47. Kteily, Bruneau, "Darker Demons of Our Nature."

48. N. S. Kteily, E. Bruneau, "Darker Demons of Our Nature: The Need to (Re) Focus Attention on Blatant Forms of Dehumanization," *Current Directions in Psychological Science* 26, 487–94 (2017).

49. P. A. Goff, J. L. Eberhardt, M. J. Williams, M. C. Jackson, "Not Yet Human: Implicit Knowledge, Historical Dehumanization, and Contemporary Consequences," *Journal of Personality and Social Psychology* 94, 292–306 (2008).

50. P. A. Goff, M. C. Jackson, B. A. L. Di Leone, C. M. Culotta, N. A. DiTomasso, "The Essence of Innocence: Consequences of Dehumanizing Black Children," *Journal of Personality and Social Psychology* 106, 526–45 (2014).

51. In social psychology, dehumanization has "attracted only scattered attention" (N. Haslam, "Dehumanization: An Integrative Re-

view," *Personality and Social Psychology Review* 10, 252–64 [2006]), and the "contributions of psychologists to the literature of dehumanization have been relatively scant" (P. A. Goff, J. L. Eberhardt, M. J. Williams, M. C. Jackson, "Not Yet Human: Implicit Knowledge, Historical Dehumanization, and Contemporary Consequences," *Journal of Personality and Social Psychology* 94, 292–306 [2008]).

52. A. Gordon, "Here's How Often ESPN Draft Analysts Use the Same Words Over and Over" in *Vice Sports* (2015), published online May 4, 2015, https://sports.vice.com/en_us/article/4x9983/heres-how-often-espn-draft-analysts-use-the-same-words-over-and-over.

53. CBS News, "Curious George Obama Shirt Causes Uproar," in CBS News (2008). Published online May 15, 2008, https://www.cbsnews.com/news/curious-george-obama-shirt-causes-uproar/.

54. S. Stein, "New York Post Chimp Cartoon Compares Stimulus Author to Dead Chimpanzee," *Huffington Post,* March 21, 2009.

55. George W. Bush was simianized on various websites with titles like "Bush Monkey Comparisons" that juxtaposed photos of Bush's facial expressions with photos of chimpanzees. But these comparisons were more commonly made to Barack Obama and his family. In 2016 a Clay County employee called Michelle Obama "a [*sic*] ape in heels" (C. Narayan, 2016). Fox News readers called Obama's daughter Malia an "ape" and a "monkey" (K. D'Onofrio, in Diversity Inc., 2016). Dan Johnson, a member of the Kentucky state house referred to the Obamas as "monkeys" (L. Smith, in WDRB, 2016). Photoshopped pictures of the Obamas with chimpanzee and gorilla faces also went viral on social media (K. B. Kahn, P. A. Goff, J. M. McMahon, "Intersections of Prejudice and Dehumanization. Simianization: Apes, Gender, Class, and Race," 6, 223 [Zurich: Verlag GmbH, 2015]).

56. J. D. Vance, *Hillbilly Elegy* (New York: HarperCollins, 2016).

57. A. Jardina, S. Piston, "Dehumanization of Black People Motivates White Support for Punitive Criminal Justice Policies," paper presented at the Annual Meeting of the American Political Science Association, September 1, 2016; Ashley Jardina, *White Identity Politics,* (Cambridge: Cambridge University Press, 2019).

58. K. M. Hoffman, S. Trawalter, J. R. Axt, M. N. Oliver, "Racial Bias in Pain Assessment and Treatment Recommendations, and False Beliefs About Biological Differences Between Blacks and Whites," *Proceedings of the National Academy of Sciences* 113, 4296–4301 (2016).

59. A. Cintron, R. S. Morrison, "Pain and Ethnicity in the United States: A Systematic Review," *Journal of Palliative Medicine* 9, 1454–73 (2006).

60. M. Peffley, J. Hurwitz, "Persuasion and Resistance: Race and the Death Penalty in America," *American Journal of Political Science* 51, 996–1012 (2007).

61. S. Ghoshray, "Capital Jury Decision Making: Looking Through the Prism of Social Conformity and Seduction to Symmetry," *University of Miami Law Review* 67, 477 (2012).

62. A. Avenanti, A. Sirigu, S. M. Aglioti, "Racial Bias Reduces Empathic Sensorimotor Resonance with Other-Race Pain," *Current Biology* 20, 1018–22 (2010).

63. N. Lajevardi, K. A. Oskooii, "Ethnicity, Politics, Old-fashioned Racism, Contemporary Islamophobia, and the Isolation of Muslim Americans in the Age of Trump," *Journal of Race, Ethnicity and Politics* 3, 112–52 (2018).

64. F. Galton, *Inquiries into Human Faculty and Its Development* (Macmillan, 1883).

65. P. A. Lombardo, *A Century of Eugenics in America: From the Indiana Experiment to the Human Genome Era* (Bloomington: Indiana University Press, 2011).

66. V. W. Martin, C. Victoria, *The Rapid Multiplication of the Unfit* (London, 1891).

67. H. Sharp, *The Sterilization of Degenerates* (1907).

68. S. Kühl, *For the Betterment of the Race: The Rise and Fall of the International Movement for Eugenics and Racial Hygiene* (Palgrave Macmillan, 2013).

69. Lyudmila Trut, "Early Canid Domestication: The Farm-Fox Experiment; Foxes Bred for Tamability in a 40-year Experiment Exhibit Remarkable Transformations That Suggest an Interplay Between Behavioral Genetics and Development," *American Scientist* 87, 160–69 (1999).

70. A. R. Wood, T. Esko, J. Yang, S. Vedantam, T. H. Pers, S. Gustafsson, A. Y. Chu, K. Estrada, J. a. Luan, Z. Kutalik, "Defining the Role of Common Variation in the Genomic and Biological Architecture of Adult Human Height," *Nature Genetics* 46, 1173–86 (2014).

71. C. F. Chabris, J. J. Lee, D. Cesarini, D. J. Benjamin, D. I. Laibson, "The Fourth Law of Behavior Genetics," *Current Directions in Psychological Science* 24, 304–12 (2015).

72. M. Lundstrom, "Moore's Law Forever?" *Science* 299, 210–11 (2003).

73. R. Kurzweil, "The Law of Accelerating Returns," in *Alan Turing: Life and Legacy of a Great Thinker* (New York: Springer, 2004), 381–416.

74. J. Dorrier, "Service Robots Will Now Assist Customers at Lowe's Stores," in *Singularity Hub* (2014).

75. J. J. Duderstadt, *The Millennium Project* (1997).

76. J. Glenn, *The Millennium Project: State of the Future* (Washington, D.C.: World Federation of U.N. Assocations, 2011).

8 The Highest Freedom

1. R. W. Wrangham, "Two Types of Aggression in Human Evolution," *Proceedings of the National Academy of Sciences* 201713611 (2017).

2. C. J. von Rueden, "Making and Unmaking Egalitarianism in Small-Scale Human Societies" *Current Opinion in Psychology* 33, 167–171 (2019).

3. There are exceptions to this—for example, the potlatch Native Americans, who had abundant natural coastal resources but were not agriculturalists, did develop forms of social hierarchy. W. Suttles, "Coping with Abundance: Foraging on the Northwest Coast," in *Man the Hunter* (Routledge, 2017), 56–68.

4. Peter Turchin, *Ultrasociety: How 10,000 Years of War Made Humans the Greatest Cooperators on Earth* (Smashwords edition, 2015, smashwords.com/books/view/593854).

5. E. Weede, "Some Simple Calculations on Democracy and War Involvement," *Journal of Peace Research* 29, 377–83 (1992).

6. J. R. Oneal, B. M. Russett, "The Kantian Peace: The Pacific Benefits of Democracy, Interdependence, and International Organizations" in *Bruce M. Russett: Pioneer in the Scientific and Normative Study of War, Peace, and Policy* (New York: Springer, 2015), 74–108.

7. C. B. Mulligan, R. Gil, X. Sala-i-Martin, "Do Democracies Have Different Public Policies than Nondemocracies?" *Journal of Economic Perspectives* 18, 51–74 (2004).

8. J. Tavares, R. Wacziarg, "How Democracy Affects Growth," *European Economic Review* 45, 1341–78 (2001).

9. M. Rosen, "Democracy," in *Our World in Data* (May 1, 2017, https://ourworldindata.org/democracy).

10. H. Hegre, "Toward a Democratic Civil Peace?: Democracy, Political Change, and Civil War," in *American Political Science Association,* vol. 95 (Cambridge University Press, 2001), 33–48.

11. H. Hegre, "Democracy and Armed Conflict," *Journal of Peace Research* 51, 159–172 (2014).

12. Peter Levine, *The New Progressive Era: Toward a Fair and Deliberative Democracy* (Lanham, MD: Rowman & Littlefield, 2000).

13. James Madison, The Federalist no. 10 (1787).

14. Thomas Paine, *Common Sense* (Penguin, 1986).

15. Cass R. Sunstein, *Can It Happen Here?: Authoritarianism in America* (New York: Dey Street Books, 2018).

16. James Madison, The Federalist no. 51 (1788).

17. James Madison, John Jay, Alexander Hamilton, *The Federalist Papers,* edited by Jim Miller (Mineola, NY: Dover Publications, 2014), 253–57.

18. R. A. Dahl, *How Democratic Is the American Constitution?* (New Haven, CT: Yale University Press, 2003).

19. Michael Ignatieff, *American Exceptionalism and Human Rights* (Princeton, NJ: Princeton University Press, 2009).

20. M. Flinders, M. Wood, "When Politics Fails: Hyper-Democracy and Hyper-Depoliticization," *New Political Science* 37, 363–81 (2015).

21. D. Amy, *Government Is Good: An Unapologetic Defense of a Vital Institution* (New York: Dog Ear Publishing, 2011).

22. A. Romano, "How Ignorant Are Americans?" *Newsweek,* March 20, 2011.

23. Annenberg Public Policy Center, University of Pennsylvania, "Americans Know Surprisingly Little About Their Government, Survey Finds," September 17, 2014.

24. A. Davis, "Racism, Birth Control and Reproductive Rights," *Feminist Postcolonial Theory—A Reader,* 353–67 (2003).

25. In November 2013, the approval rate of Congress fell to 9 percent—the lowest it has been since the Gallup poll started measuring its approval in 1974. It barely improved in the next few years, with 2016 showing an approval rate of only 13 percent (Gallup, "Congress and the Public," 2016). We usually have more faith in the Supreme Court and the presidency, but a 2014 poll showed approval of the Supreme Court at a record low of 30 percent and the presidency at a six-year low of 35 percent (J. McCarthy, "Americans Losing Confidence in All Branches of US Gov't.," Gallup, 2014).

26. R. S. Foa, Y. Mounk, "The Democratic Disconnect," *Journal of Democracy* 27, 5–17 (2016).

27. One survey found that only 32 percent of millennials consider it essential to live in a democracy; one in four thought that democracy was a "bad" or "very bad" way to run the country, and more than 70 percent thought that it is legitimate for the military to take over when the government is failing at its job. And these numbers have been getting worse. In 1990, 53 percent of young Americans were interested in politics. In 2005, the number had dropped to 41 percent. In 1995, only 1 in 16 Americans thought it would be good for "the army to rule"; by 2005, it was up to 1 in 6. From 1995 to 2011 there was an increase from 25 percent to 36 percent of Americans who favored a "strong leader who doesn't have to bother with parliament and elections." By 2011, it was up to a third. R. S. Foa, Y. Mounk, "The Democratic Disconnect," *Journal of Democracy* 27, 5–17 (2016).

28. W. Churchill, speech to the British House of Commons on November 11, 1947, https://api.parliament.uk/historic-hansard/commons/1947/nov/11/parliament-bill#S5CV0444P0_19471111_HOC_292.

29. A. Sullivan, "Democracies End When They Are Too Democratic," *New York Magazine,* May 1, 2016.

30. F. M. Cornford, ed., *"The Republic" of Plato,* vol. 30 (London: Oxford University Press, 1945).

31. J. Duckitt, "Differential Effects of Right Wing Authoritarianism and Social Dominance Orientation on Outgroup Attitudes and Their Mediation by Threat from and Competitiveness to Outgroups," *Personality and Social Psychology Bulletin* 32, 684–96 (2006).

32. A. K. Ho, J. Sidanius, N. Kteily, J. Sheehy-Skeffington, F. Pratto, K. E. Henkel, R. Foels, A. L. Stewart, "The Nature of Social Dominance Orientation: Theorizing and Measuring Preferences for Intergroup Inequality Using the New SDO_7 Scale," *Journal of Personality and Social Psychology* 109, 1003 (2015).

33. Previously perceived as small disparate fringe groups, as of 2016, the alt-right has been characterized as a single group and has begun displaying considerable influence in American politics. Internationally, in July 2016, thirty-nine countries in Europe had alt-right parties in their parliament. Marine Le Pen's National Front, which was founded by her neo-Nazi father, got all the way to the final round of the presidential elections in France. Frauke Petry's Alternative for Germany, who suggested shooting asylum seekers as they crossed the

border and banning Islamic symbols, is the country's third most powerful political party. Pegida, the German anti-Islamic party, is gaining the kind of support reminiscent of a Trump rally, where journalists need security guards. Geert Wilders, who was charged with inciting violence against Muslims, currently heads the most popular party in the Netherlands. If the election had been held in 2016, he would have won more seats in the government than any other party. The party has called to close Islamic schools and record the ethnicity of Dutch citizens. They would like to deport foreign criminals, abolish the senate, and leave the EU. He has compared the Quran to Hitler's autobiography and funded anti-Islamic propaganda. Golden Dawn, with its symbol that looks suspiciously like a swastika and whose members give Nazi salutes, is the third-largest party in Greece. Jobbik, the third-largest party in Hungary, has issued the kind of anti-Semetic rhetoric that would make Hitler proud. The Sweden Democrats—a misleading name, since it was founded by white supremacists—is the third-largest party in Sweden. The Freedom Party, which was founded by an SS officer, lost the 2016 Austrian presidential elections by a whisker (Daniel Koehler, "Right-Wing Extremism and Terrorism in Europe: Current Developments and Issues for the Future," *PRISM: The Journal of Complex Operations,* National Defense University 6, no. 2, July 18, 2016).

34. Koehler, "Right-Wing Extremism and Terrorism in Europe."

35. P. S. Forscher, N. Kteily, "A Psychological Profile of the Alt-right," *Perspectives in Psychological Science* doi .org/10.1177/1745691619868208 (2019).

36. J. D. Vance, *Hillbilly Elegy* (New York: HarperCollins, 2016).

37. Karen Stenner, *The Authoritarian Dynamic* (Cambridge: Cambridge University Press, 2005).

38. K. Costello, G. Hodson, "Lay Beliefs About the Causes of and Solutions to Dehumanization and Prejudice: Do Nonexperts Recognize the Role of Human–Animal Relations?" *Journal of Applied Social Psychology* 44, 278–88 (2014).

39. E. L. Paluck, D. P. Green, "Prejudice Reduction: What Works? A Review and Assessment of Research and Practice," *Annual Review of Psychology* 60, 339–67 (2009); Robin Diangelo, *White Fragility: Why It's So Hard for White People to Talk About Racism* (Boston: Beacon Press, 2018).

40. P. Henry, J. L. Napier, "Education Is Related to Greater Ideological Prejudice," *Public Opinion Quarterly* 81, 930–42 (2017).

41. Ashley Jardina, *White Identity Politics* (Cambridge: Cambridge University Press, 2019).

42. G. M. Gilbert, *The Psychology of Dictatorship: Based on an Examination of the Leaders of Nazi Germany* (New York: Ronald Press Company, 1950).

43. Walter Sinnott-Armstrong, *Think Again: How to Reason and Argue* (Oxford University Press, 2018).

44. C. Andris, D. Lee, M. J. Hamilton, M. Martino, C. E. Gunning, J. A. Selden, "The Rise of Partisanship and Super-cooperators in the US House of Representatives," *PLoS One* 10, e0123507 (2015).

45. T. E. Mann, N. J. Ornstein, *It's Even Worse Than It Looks: How the American Constitutional System Collided with the New Politics of Extremism* (New York: Basic Books, 2016).

46. Mike Lofgren, *The Party Is Over: How Republicans Went Crazy, Democrats Became Useless, and the Middle Class Got Shafted* (New York: Viking Penguin, 2012).

47. K. Gehl, M. E. Porter, *Why Competition in the Politics Industry Is Failing America: A Strategy for Reinvigorating Our Democracy,* Harvard Business School paper, September 2017, www.hbs.edu/ competitiveness/Documents/why-competition-in-the-politics -industry-is-failing-america.pdf.

48. Richard Wrangham, *The Goodness Paradox: The Strange Relationship Between Virtue and Violence in Human Evolution* (New York: Pantheon, 2019).

49. P. M. Oliner, *Saving the Forsaken: Religious Culture and the Rescue of Jews in Nazi Europe* (New Haven, CT: Yale University Press, 2008).

50. Website: "About the Righteous" (Yad Vashem: The World Holocaust Memorial Center), (2017). Published online, retrieved August 19, 2017, http://www.yadvashem.org/righteous/about-the-righteous.

51. Z. N. Hurston, August 11, 1955, letter to the *Orlando Sentinel.* Retrieved from http://teachingamericanhistory.org/library/document/ letter-to-the-orlando-sentinel/.

52. Thomas F. Pettigrew, Linda R. Tropp, *When Groups Meet: The Dynamics of Intergroup Contact* (New York: Psychology Press, 2013).

53. W.E.B. Du Bois, "Does the Negro Need Separate Schools?" *Journal of Negro Education* 4, 328–35 (1935).

54. J. W. Jackson, "Contact Theory of Intergroup Hostility: A Review and Evaluation of the Theoretical and Empirical Literature," *International Journal of Group Tensions* 23, 43–65 (1993).

55. S. E. Gaither, S. R. Sommers, "Living with an Other-race Roommate Shapes Whites' Behavior in Subsequent Diverse Settings," *Journal of Experimental Social Psychology* 49, 272–76 (2013).
56. P. B. Wood, N. Sonleitner, "The Effect of Childhood Interracial Contact on Adult Antiblack Prejudice," *International Journal of Intercultural Relations* 20, 1–17 (1996).
57. C. Van Laar, S. Levin, S. Sinclair, J. Sidanius, "The Effect of University Roommate Contact on Ethnic Attitudes and Behavior," *Journal of Experimental Social Psychology* 41, 329–45 (2005).
58. Another college study randomly assigned students to either a mixed-race or same-race roommate. Researchers surveyed the freshmen at the beginning and end of their first semester. They found that freshmen living with same race roommates reported higher initial satisfaction but this satisfaction decreased while satisfaction between mixed-race roommates did not. Meanwhile, mixed-race roommates showed significant increase in their tolerance for students of other races while same-race roommates did not. White freshmen with black roommates reported higher comfort levels interacting with other minority students by the end of the quarter than same-race roommates. At the end of the semester, an implicit measure showed that mixed-race roommates showed increased positivity toward black people while participants with same-race roommates did not.
N. J. Shook, R. H. Fazio, "Interracial Roommate Relationships: An Experimental Field Test of the Contact Hypothesis," *Psychological Science* 19, 717–23 (2008).
59. D. M. Wilner, R. P. Walkley, S. W. Cook, *Human Relations in Interracial Housing* (Minneapolis: University of Minnesota Press, 1955).
60. J. Nai, J. Narayanan, I. Hernandez, K. J. Savani, "People in More Racially Diverse Neighborhoods Are More Prosocial," *Journal of Personality and Social Psychology* 114, 497 (2018).
61. R. Falvo, D. Capozza, G. A. Di Bernardo, A. F. Pagani, "Can Imagined Contact Favor the 'Humanization' of the Homeless?" *TPM: Testing, Psychometrics, Methodology in Applied Psychology* 22 (2015).
62. L. Vezzali, M. D. Birtel, G. A. Di Bernardo, S. Stathi, R. J. Crisp, A. Cadamuro, E. P. Visintin, "Don't Hurt my Outgroup Friend: Imagined Contact Promotes Intentions to Counteract Bullying," *Group Processes & Intergroup Relations* (2019).
63. D. Broockman, J. Kalla, "Durably Reducing Transphobia: A Field

Experiment on Door-to-Door Canvassing," *Science* 352, 220–24 (2016).

64. D. Capozza, G. A. Di Bernardo, R. Falvo, "Intergroup Contact and Outgroup Humanization: Is the Causal Relationship Uni- or Bidirectional?" *PLoS One* 12, e0170554 (2017).

65. G. Hodson, "Do Ideologically Intolerant People Benefit from Intergroup Contact?" *Current Directions in Psychological Science* 20, 154–59 (2011).

66. G. Hodson, R. J. Crisp, R. Meleady, M.J.P.o.P.S. Earle, "Intergroup Contact as an Agent of Cognitive Liberalization," *Perspectives on Psychological Science* 13, 523–48 (2018).

67. B. Major, A. Blodorn, G. Major Blascovich, "The Threat of Increasing Diversity: Why Many White Americans Support Trump in the 2016 Presidential Election," *Group Processes & Intergroup Relations* 21, 931–40 (2018).

68. F. Beyer, T. F. Münte, C. Erdmann, U. M. Krämer, "Emotional Reactivity to Threat Modulates Activity in Mentalizing Network During Aggression," *Social Cognitive and Affective Neuroscience* 9, 1552–60 (2013).

69. N. Kteily, G. Hodson, E. Bruneau, "They See Us as Less Than Human: Metadehumanization Predicts Intergroup Conflict via Reciprocal Dehumanization," *Journal of Personality and Social Psychology* 110, 343 (2016).

70. Contact has been demonstrated to increase tolerance toward ethnic groups such as Chinese students, black workers in South Africa, Turkish schoolchildren in Germany, and Southeast Asian immigrants in Australia. It works toward those who are traditionally dehumanized, like the elderly, the mentally ill, people with AIDS, disabled people, and even computer programmers. T. F. Pettigrew, "Intergroup Contact Theory," *Annual Review of Psychology* 49, 65–85 (1998). We could not find any evidence of where contact has systematically or repeatedly failed to improve social relationships between groups.

71. Some have argued against this idea by suggesting that political boundaries separating populations are the most effective means to maintain peace. A. Rutherford, D. Harmon, J. Werfel, A. S. Gard-Murray, S. Bar-Yam, A. Gros, R. Xulvi-Brunet, Y. Bar-Yam, "Good Fences: The Importance of Setting Boundaries for Peaceful Coexistence," *PLoS One* 9, e95660 (2014). Others have argued the contact hypothesis has not yet been fully evaluated. While dozens of

studies show consistent effects, they are often moderate and most only evaluate prejudice reduction and not dehumanization (that has not been typically been measured). E. L. Paluck, S. A. Green, D. P. Green, "The Contact Hypothesis Re-evaluated," *Behavioural Public Policy* 3, 129-158 (2019).

72. N. Haslam, S. J. Loughnan, "Dehumanization and Infrahumanization," *Annual Review of Psychology* 65, 399–423 (2014).

73. "American Values Survey 2013," Public Religion Research Institute, retrieved October 18, 2017, from https://www.prri.org/wp -content/uploads/2014/08/AVS-Topline-FINAL.pdf.

74. T. W. Smith, P. Marsden, M. Hout, J. Kim, "General Social Surveys, 1972–2016," NORC at the University of Chicago, 2016.

75. M. Saincome, "Berkeley Riots: How Free Speech Debate Launched Violent Campus Showdown," *Rolling Stone,* February 6, 2016.

76. Frantz Fanon, *The Wretched of the Earth,* translated by Constance Farrington, with a preface by Jean-Paul Sartre (New York: Grove Press, 1963), vol. 36.

77. J. Lyall, I. Wilson, "Rage Against the Machines: Explaining Outcomes in Counterinsurgency Wars," *International Organization* 63, 67–106 (2009).

78. Malcolm X with Alex Haley, *The Autobiography of Malcolm X* (New York: Grove Press, 1965).

79. E. Chenoweth "The Success of Nonviolent Civil Resistance" in TedX Boulder (2013). Published online, https://www.youtube.com/ watch?v=YJSehRlU34w.

80. E. Chenoweth, M. J. Stephan, *Why Civil Resistance Works: The Strategic Logic of Nonviolent Conflict* (New York: Columbia University Press, 2011).

81. M. Feinberg, R. Willer, C. Kovacheff, "Extreme Protest Tactics Reduce Popular Support for Social Movements," *Rotman School of Management Working Paper 2911177* (2017); B. Simpson, R. Willer, M. Feinberg, "Does Violent Protest Backfire?: Testing a Theory of Public Reactions to Activist Violence," *Socius: Sociological Research for a Dynamic World* 4, 2018, doi.org/10.1177/2378023118803189.

82. E. Volokh, *The First Amendment and Related Statutes* (New York: Foundation Press, 2011).

83. Samuel Walker, *Hate Speech: The History of an American Controversy* (Lincoln: University of Nebraska Press, 1994).

84. Toni M. Massaro, "Equality and Freedom of Expression: The Hate Speech Dilemma," *William & Mary Law Review* 32 (1991), https://scholarship.law.wm.edu/wmlr/vol32/iss2/3.

85. D. Meagher, "So Far So Good: A Critical Evaluation of Racial Vilification Laws in Australia," *Federal Law Review* 32 (2004), 225.

86. M. Bohlander, *The German Criminal Code: A Modern English Translation* (New York: Bloomsbury, 2008).

87. A. Gow, "'I Had No Idea Such People Were in America!': Cultural Dissemination, Ethno-linguistic Identity and Narratives of Disappearance," spacesofidentity.net 6 (2006).

88. E. Bruneau, N. Jacoby, N. Kteily, R. Saxe, "Denying Humanity: The Distinct Neural Correlates of Blatant Dehumanization," *Journal of Experimental Psychology: General* 147, 1078–1093 (2018).

89. N. L. Canepa, "From Court to Forest: The Literary Itineraries of Giambattista Basile," *Italica* 71, 291–310 (1994).

90. C. Johnson, "Donald Trump Says the US Military Will Commit War Crimes for Him," Fox News Debate, published online March 4, 2016. https://www.youtube.com/watch?time_continue=9&v=u3LszO-YLa8.

91. B. Kentish, "Donald Trump Blames 'Animals' Supporting Hillary Clinton for Office Firebomb Attack," *The Independent* (2016), published online October 17, 2016, http://www.independent.co.uk/news/world/americas/us-elections/us-election-donald-trump-hillary-clinton-animals-firebomb-attack-north-carolina-republican-party-a7365206.html.

92. M. Miller, "Donald Trump On a Protester: 'I'd Like to Punch Him in the Face,'" *Washington Post* (2016). Published online February 23, 2016, https://www.washingtonpost.com/news/morning-mix/wp/2016/02/23/donald-trump-on-protester-id-like-to-punch-him-in-the-face/.

93. J. Diamond, "Trump: I Could Shoot Somebody and Not Lose Voters" *CNN Politics* (2016). Published online January 24, 2016, http://www.cnn.com/2016/01/23/politics/donald-trump-shoot-somebody-support/.

94. Jane Jacobs, *The Death and Life of Great American Cities* (New York: Vintage, 2016).

95. Richard Florida, *The New Urban Crisis: How Our Cities Are Increasing Inequality, Deepening Segregation, and Failing the Middle Class, and What We Can Do About It* (UK: Hachette, 2017).

96. R.T.T. Forman, "The Urban Region: Natural Systems in Our Place, Our Nourishment, Our Home Range, Our Future," *Landscape Ecology* 23 (2008), 251–53.

97. A. Andreou, "Anti-Homeless Spikes: Sleeping Rough Opened My Eyes to the City's Barbed Cruelty," *Guardian* 19 (2015), 4–8.

9 Circle of Friends

1. R. M. Beatson, M. J. Halloran, "Humans Rule! The Effects of Creatureliness Reminders, Mortality Salience and Self-esteem on Attitudes Towards Animals," *British Journal of Social Psychology* 46, 619–32 (2007).

2. K. Costello, G. Hodson, "Lay Beliefs about the Causes of and Solutions to Dehumanization and Prejudice: Do Nonexperts Recognize the Role of Human-Animal Relations?" *Journal of Applied Social Psychology* 44, 278–88 (2014).

3. K. Dhont, G. Hodson, K. Costello, C. C. MacInnis, "Social Dominance Orientation Connects Prejudicial Human–Human and Human–Animal Relations," *Personality and Individual Differences* 61, 105–108 (2014).

4. K. Costello, G. Hodson, "Exploring the Roots of Dehumanization: The Role of Animal-Human Similarity in Promoting Immigrant Humanization," *Group Processes & Intergroup Relations* 13, 3–22 (2010).

5. R. B. Bird, D. W. Bird, B. F. Codding, C. H. Parker, J. H. Jones, "The 'Fire Stick Farming' Hypothesis: Australian Aboriginal Foraging Strategies, Biodiversity, and Anthropogenic Fire Mosaics," *Proceedings of the National Academy of Sciences* 105, 14796–801 (2008).

6. H. G. Parker, L. V. Kim, N. B. Sutter, S. Carlson, T. D. Lorentzen, T. B. Malek, G. S. Johnson, H. B. DeFrance, E. A. Ostrander, L. Kruglyak, "Genetic Structure of the Purebred Domestic Dog," *Science* 304, 1160–64 (2004).

7. H.J.V.S. Ritvo, "Pride and Pedigree: The Evolution of the Victorian Dog Fancy," *Victorian Studies* 29, 227–253 (1986).

8. Michael Worboys, Julie-Marie Strange, Neil Pemberton, *The Invention of the Modern Dog* (Baltimore: Johns Hopkins University Press, 2019).

Image Credits

73   Da Vinci, Leonardo. *Mona Lisa*. 1503–1506, Louvre; Creative Commons

87   Pearson, Karl. A monograph on albinism in man. Vol. 6. Dulau, 1913. Plate BB; Creative Commons

124   Photographer unknown, Ota Benga. 1905–1906, Library of Congress; Creative Commons

128, left   *King Kong* movie poster, RKO Radio Pictures, 1933; Creative Commons

128, right   Khan, Lin Shi, and Tony Perez, *Scottsboro, Alabama: A Story in Linoleum Cuts,* NYU Press, 2003; Creative Commons

137   Kimball, Linda, porch monkey rant, Facebook, http://41af3k34gprx4f6bg12df75i.wpengine.netdna-cdn.com/wp -content/uploads/sites/19/2017/10/ORIG-POST-ALONE.jpeg; Facebook

173   Unknown, Eisenhower at Ordurf, 1945; National Archives and Records Administration, College Park

175   Pakhrin, S., Women's March, 2017; Wiki Commons

Index

Page numbers of illustrations and photographs appear in italics.

Ache people, 144
aggression
  amygdala and, 51, 56, 57, 96,
    112–14
  appearing young, as protective,
    79–80
  beneficial contact to lower,
    170–71, 249–50n71
  bonobos and, 46–47, 49, 50, 51,
    54, 93, 109, 217n17
  cavefish and, 227–28n22
  chimpanzees and, 39–42, 50, 93
  as costly to a species, xvii, 42
  in crows, 79
  dehumanization and, 111, 180,
    181
  democracies and, 152
  in dogs, 109, 190, 195
  domestication and, 85, 109, 110
  Dutch men scenario, 113–14
  Ecstasy aftermath and, 71
  finger length (2D:4D ratio)
    and, 51
  friendliness more advantageous
    than, 42, 46–47, 80, 98
  friendly foxes and, 25, 146

  hunter-gatherer cultures and, 93
  oxytocin and, 109–10, 113,
    232n5
  serotonin and, 25, 51, 71, 110,
    227–28n22
  testosterone and, 50–51,
    67–68, 70
  threat from outsiders or strangers
    and, 109–12, 114–20, 232n5
agricultural communities, 150–51
Allport, Gordon, 132–33
alt-right, 156–61, 176, 245–46n33
  aggression, intolerance, and,
    157–58
  conformity, group identity, and,
    157
  defined, 156
  dehumanization by, *158*, 158–59,
    159
  "hate speech" and, 179
  survival of the fittest and, 156–57
  white supremacy, 158–59, *159*,
    163, 178
Alves, Dani, 138
American Civil Liberties Union
  (ACLU), 178–79

Americans for Tax Reform, 155
Andersson, Leif, 22
André, Claudine, 43, 186–89, *188*
Andreou, Alex, 184
animals, 66
  animal-human divide, 190–91
  brain size and, 63, 70
  cooperative communication in,
    xxiv–xxv, 6, 10–14, 15, 16, 27,
    31–32, 49, 51–56, 84,
    203–4n8
  domestication, aggression and,
    85, 109, 110
  domestication, juvenile
    development and, 78–92,
    203–4n8
  domestication, physical and
    physiological traits related to
    (domestication syndrome), 8,
    21–22, 24–25, 31, 36, 55–58,
    66, 70, 203–4n8
  genetic changes in, 84
  intelligence and, 8, 36, 55–58,
    203–4n8
  judging on physical traits,
    189–90
  kindness and, 188–91
  marshmallow test for, 60–61
  self-control and, 60
  serotonin levels and, 25, 26,
    51, 70
  theory of mind and, 6
  urban, self-domestication and,
    35–36
  See also *specific species*
antifa, 163, 174–76
anti-Semitism, 159–60, 165–66,
  245–46n33
ants as superorganisms, xviii
apes, great apes, 35, 61, 88, *124*,
  124–25, 126, 127–28, 241n55
  King Kong, 127–28, *128*, 138
  Linnaeus's classification, 126,
    127

  on nineteenth century
    evolutionary ladder, 126–27
  the uncanny valley and, 125–26
  See also simianization; *specific
    types of apes*
Aronson, Elliot, xiv–xv
Ascent of Man Scale, 116–17, *117*,
  118, 142–43
Asch, Solomon, 132–33
  conformity experiment, 132–33,
    133
Astyanax (blind cavefish),
  227–28n22
"Attitude, Inequality Mismatch"
  (Goff), 130–31
Australia
  Aboriginals of, 64–65, 191–93
  hate speech laws, 180
australopithecine, 119
axolotl salamander, 78, 84

Baka pygmies, 122
Bandura, Albert, 134–36
Banyamulenge tribe, 102–8
  Gatumba Massacre, Hutu
    attacks on, 107–8, 109
  Rachel's story, 102–8, *103*,
    108
Basile, Giambattista, 181
behavioral modernity, xxiv, *64*, 67,
  222n353
Belgium, 104–5
Belyaev, Dmitry, 18–19, 79
  domestication of foxes
    experiment, 19, 23–27, *24*,
    39, 55–56
Belyaev, Nikolai, 18
beneficial contact, 165–74, *170*
  aggression lowered by, 170–71,
    249–50n71
  in education, xv–xvi, 167, *168*
  Eisenhower's granddaughter's
    anecdote, 172–74
  in housing, 169, 248n58

imaginary characters or virtual
contact and, 169
in the military, 168
true friendship, 171–72
World War II, with Jews,
165–66, 171
Benga, Ota, 123
Bengalese finches, 36–37, 52, 99
Biden, Joe, xxix
Bird, Doug, 192
birds, 36–37, 52, 99
blackbirds, urban, 35–36
Blake, William, 91
blue streak wrasse, 80
Boeckx, Cedric, 87
bonobos, xix, 42–54, 47, 99,
228n26
aggression and, 46–47, 49, 50,
51, 54, 93, 109, 217n17
amygdala of, 51
attraction to strangers, 50, 93
babies, status of, 44, 45, 48
brain size, 39, 63
chimpanzees vs., 39, 49, 52
comparison of human, bonobo,
and chimpanzee babies,
88–89
cooperative communication and,
49, 51–54, 53
cortisol levels in, 50
dominance equation and, 44
evolution of cognition vs.
humans, 59
eyes of, 73
female bonding, 45–46, 109
female ovulation, mating, 44–45,
46, 47, 82
friendliness of, 38, 46–47, 50
gaze direction and, 51–52,
74–75, 218n34
genetic changes in, 84
juvenile behaviors and, 81, 82
Malou (rescued bonobo),
42–44, 48

self-domestication and, xxvi,
48–51
serotonin in, 51
sexual behaviors of, 82–83
sharing food and, 53
social interactions and, 46, 66
social networks, 93
testing tolerance in, 49–50
testosterone in, 82
thyroid hormone in, 227n20
traits of, 38–39, 49, 72
vocal flexibility, 52, 218n35
Boston Marathon bombing, 235n38
brain, 58
amygdala and threat response,
51, 56, 57, 96, 112–14
animal, development of, 91
capybara vs. rhesus macaque, 63
dehumanization and, 112–14,
119
domestication and size of, 70
of hominins, 64
human, xix, 63, 64, 66, 70, 88,
90–91, 99, 119
hypothalamic pituitary gonadal
(HPG) axis, 86
medial prefrontal cortex
(mPFC), 57, 96–97, 112, 114
neural crest cells and, 86
neurons responding to eyes, 75
oxytocin and, 112, 113
parietal regions, 90
precuneus (PC), 57, 90, 112, 119
prefrontal cortex (PFC), 59
reproductive cycles and, 86
self control and, 59, 61–62
superior temporal sulcus (STS),
75, 112, 114
synaptic pruning, 90
temporal parietal junction (TPJ),
57, 90, 112, 114
theory of mind network, 57, 75,
90, 110, 180
See also cognition

Brooks, James, 35
Bündchen, Gisele, 138
Burnham, Terry, 75
Burundi, 106, 107
Bush, George H. W., 121
Bush, George W., 241n55

Candid Critters project, 214n43
capybara, 63
Ceiri, Bob, 69
Chenoweth, Erica, 176–78, 177, 178
chickens, 26–27
children, babies
    communicative intention skills
        (pointing, gestures), 3–4, 6–7,
        14n, 15, 32, 51
    comparison of human, bonobo,
        and chimpanzee babies,
        88–89
    emotional reactivity, as
        predictive, 56
    eye contact and gaze direction,
        74–75
    prejudice against black children,
        141–42
    theory of mind in, 56
    vocal direction and, 14
chimpanzees, xix, 11, 47
    aggression and, 39–42, 50, 93
    bonobos vs., 39, 49, 52
    brain development and, 91
    brain size vs. humans, 63
    cooperation among, 46, 53, 93
    cooperative communication and,
        xxiv–xxv, 6, 7, 13, 14–16,
        51–54
    eye contact and, 15, 218n34
    eyes of, 73, 73
    female hierarchies, 41
    female ovulation and mating,
        45, 47
    gaze direction and, 51–52,
        74–75

hostility to strangers, 93
    infant chimpanzees, 44, 45,
        88–89
    infanticide among, 45
    male dominance, 47
    as mirror of human behavior, 54
    as primed for competition, 50
    sign of trust among, 40
    testing tolerance in, 49–50
    testosterone levels in, 50–51
    theory of mind and, 6
    thyroid hormone in, 227n20
    vocal flexibility, 52
Chirac, Jacques, 43
Churchill, Steve, 69
Churchill, Winston, 156
Citalopram, 71
Clay, Zanna, 218n35
Clinton, Hillary, 181
cockroaches, 79
cognition
    brain size and, 62
    comparison of human, bonobo,
        and chimpanzee babies,
        88–89
    cooperation and, 80
    domestication and, 22, 27,
        28–31, 30, 36, 37, 55, 70,
        203–4n8
    in dogs, 8, 10, 16, 28, 36
    evolution of, bonobos vs.
        humans, 59
    in foxes, 27–28, 31
    friendliness and, xxv, 98
    human, xxiii–xxiv, xxv, 16, 59,
        63–64, 88, 91, 203–4n8
    human-specific adaptation and
        social cognition, 66, 90, 91
    intelligence, undue emphasis on,
        196
    natural selection and, 58
    self-control and, 59–63
    social tolerance and, 98
    temperament and, 58

competition, xiv, xxv, 6, 46, 50–51, 57, 65, 68, 113, 158, 196, 232n5
Congo Wars of 1996 and 1998, 105–6
    André and bonobo protection, 186–89
    Tutsis and, 186
Cook, Tim, 148
cooperation
    beneficial contact and, 168–71, 174
    in bonobos and chimpanzees, 52–54, 53, 93
    cognitive changes and, 98, 100
    Darwin's observations and, xvii
    in education, xv–xvi, 168
    emotional reactivity and, 57
    as evolutionary advantage, xvii
    flowers and pollinating insects, xviii
    friendliness defined as, xvii
    hunter-gatherers and, 92–93
    group members and, 143, 232n5
    as key to species survival, xvi
    microbes and your body, xvii–xviii
    in politics, xxviii-xxix, xxx
    theory of mind and, 4, 114
    white sclerae and, 76
cooperative communication, xxiv–xxv
    in babies, children, 3–4, 14, 16, 27, 51, 88, 90, 91–92, 95
    in bonobos, 49, 51–54, 53
    in chimpanzees, xxiv–xxv, 6, 7, 13, 14–16, 51–54
    in dogs, 7, 10–14, 15, 16, 27, 31–32, 34, 212–13n28
    domestication and, 31, 203–4n8
    emotional reactivity and, 56, 58, 66
    eyes and, 73–74, 75–76
    in friendly foxes, 27, 31, 55–56, 84
    in friendly humans, 69, 78
    human survival and, xxiv–xxv
    pointing as the star of, 7
    technology development and, 65–66
    theory of mind and, 4–5
    tolerance and, 54
    See also pointing
Coulter, Ann, 179
coyotes, 35
crows, 79
cultures
    Allport's theory of prejudice, 132
    Asch's conformity experiment and, 132–33, 133
    cultural identity, 119, 166
    explanations for atrocities in, 132–37
    genocide and, 129, 130, 239–40n40
    human desire to conform and, 132
    incarceration rate in the U.S. and, 131
    Milgram's experiments on obedience to authority, 133–34
    new prejudice in postracial societies, 130–40, 239–40n40
    racism and, 129–30, 132
    rise of, xix, xxiii–xxiv, xxvi
    simianization and, 129
    See also groups

Darwin, Charles, xvii
    Descent of Man, 126
    natural selection and, xvii
    Origin of Species, xvi–xvii, 20
    survival of the fittest, meaning of, xvi
    The Variations of Plants and Animals Under Domestication, 20–21

Daschle, Tom, xxvii
deer, 36
    domestication syndrome in, 36
    piebald and albino, 36
    urban, 36
dehumanization, xxvi, xxvii,
        110–14, *117*, 117–20, *118*
    aggression, violence, and, 111,
        180, 181
    by the alt-right, *158*, 158–59
    atrocities and, 134–36
    attitude toward animals and,
        190–91
    Bandura's experiments, 134–36
    beneficial contact to stop,
        165–71, 249–50n71
    cultural norms against, 180–82
    danger to democracy and,
        155–65, 176, 180
    of enemies and, 120–21, 127
    eugenics and, 144–45
    facial perception and, 236n43
    factors driving, 189–90
    genocide and, 111, *115*, 115–16
    Goff's research on treatment of
        U.S. blacks and, 136–37
    governments co-opted for, and
        violence, 163–64
    hate speech and, 180
    human brain network for theory
        of mind and, 112–14, 119
    human self-domestication
        hypothesis prediction of, 161,
        162–63
    as human universal, 161, 163,
        165
    initiating the cycle of, 164
    of the Irish, 127
    of the Japanese, 127
    Jardina's survey and
        simianization of black
        Americans, 139–40
    March of Progress scale and,
        116, *117*, *158*, *159*

    of Muslims, 117–18, 120,
        142–43, 159, 235n38,
        245–46n33
    "mutual animosities" and, 153
    Palestinians and Israelis, 119,
        120, 142
    in polarized U.S. Congress, xxvii
    racial prejudice in America and,
        129–30, 136–42, *137*,
        239–40n40, 240n42, 241n55
    reciprocal dehumanization, 120,
        142–43, 170, 176
    reducing, ineffective tactics for,
        160–61, 167
    of Roma (Gypsies), 118–19, 159
    simianization and, 126, 127–29,
        128, *137*, 137–40, 141–42,
        241n55
    slavery justified and, 127
    social media and, xxvii
    in social psychology, 240–41n51
    of strangers, rival groups,
        outsiders, 110–12, *117*,
        117–20, 143
    tendency prejudice and, 111
    universal preferences and,
        110–12
democracy or constitutional
        democracy, 151–56
    all humans created equal and,
        182
    America as a republic, 154
    American Founders, 153–54
    beneficial contact and, 165–74
    characteristics and advantages,
        152
    Churchill on, 156
    critics of the great American
        experiment, 154–55
    Degree of Structural Inequality,
        151
    distribution of power and, 161
    free speech and, 179–82
    hyperdemocracy, 156

loss of faith in, 245n27
moderate middle vs. extremists,
    161–63, *162*
nonviolent vs. violent protests in,
    174–79, *177, 178*
protection of minorities in, 154
threats to, 155–65, 176,
    245–46n33
U.S. Constitution and, 154
U.S. political system and, 182
when democracies fail, 156
Descent of Man (Darwin), 126
development or patterns of growth
    axolotl salamander, 78
    cockroaches and termites, 79
    comparison of human, bonobo,
        and chimpanzee babies, 88–89
    crows, 79
    dogs, 81
    domestication and, 78–92,
        203–4n8
    foxes, 81
    genes and, 82–84
    human, progress of, 88–89
    juvenile behavior, appearance,
        79–80
    neural crest cells and, 84–87, *86,*
        87, 228n26
    oxytocin and, 80
    serotonin and, 80, 88,
        227–28n22
    social behavior and, 78–79,
        82–83, 89, 92–96, 203–4n8
Dhont, Kristof, 190
Diamond, Jared, 22, 23
dingoes, 191–93
dogs
    aggression and, 109, 190, 195
    barking, 81
    cognition in, 8, 10, 16, 28, 36
    cooperative communication,
        pointing gesture and, 7,
        10–14, 15, 16, 27, 31–32, 34,
        212–13n28

dog SDO survey, 194–95
domestication and, 8, 16–17,
    23, 66
European breeds, 193–94
eye contact and, 212–13n28
first dog shows, 193
genetic changes in, 17, 84
human-animal bonds and, 97,
    191–96, 212–13n28
judging on physical traits,
    189–90
juvenile behaviors, socialization,
    and, 81
Miklosi's study of pups,
    212–13n28
natural selection for
    friendliness, 81
neural crest cells and, 228n26
Oreo (author's dog), 5, 8–14, *11*
physical traits related to
    domestication (domestication
    syndrome), 17, 21–22, 25,
    34, 49
similarity to humans, 15,
    16–17, 32
success as a species, 34–35
in West Africa, 17
why they eat poop, 33–34
wolf ancestors, xxvi, 17, 23,
    33–35
domestication, 20, *20–21,* 73
    adrenal glands and, 85
    animals domesticated by
        humans, 22–23
    Belyaev's criterion for, 23, *24*
    Belyaev's experiment replicated
        with chickens, 26–27
    Belyaev's experiment with foxes,
        19, 23–27, *24,* 39, 55–56
    brain size and, 70
    canine cognition and, 8, 10, 16,
        28, 36
    cognition and, 27, 28–31, *30,* 36,
        37, 203–4n8

domestication (*cont'd*):
  communicative abilities and,
    51–54, 203–4n8
  corticosteroid (or cortisol) levels
    and, 25, 37, 50
  Darwin on, 20–21
  development, patterns of growth
    and, 78–92, 203–4n8
  Diamond's criteria for, 23
  of dogs, 8, 21–22, 25, 33–35
  domestication syndrome
    (physical and physiological
    traits related to), 8, 21–22,
    24–25, 31, 36, 55–58, 66,
    70, 72–77, 85, 87, 87,
    203–4n8
  dominant theory of, 23
  false assumptions about, 8, 22
  formula for, 27
  fossil evidence of, 67
  friendliness and, xxv–xxvi,
    23–27, 24, 33–35, 51, 54,
    89–92, 203–4n8
  neural crest cells and, 84–87, 86,
    87, 228n26
  reproductive cycles and, 22,
    25, 27
  serotonin levels and, 25, 26, 51
  See also human self-
    domestication hypothesis;
    self-domestication; *specific*
    animals
Du Bois, W.E.B., 167

Ecstasy (drug), 71
education
  beneficial contact to reduce
    intolerance, 168
  competitive model, xiv, xxx
  cooperative learning, xv–xvi
  example, Carlos in Austin, Texas,
    xiii–xiv, 168
  "jigsaw" method, xv–xvi, 168
  school desegregation, xiii–xiv

  segregation vs. integration, 167,
    168
Eichmann, Adolf, 134
Eisenhower, Dwight D., 172–74,
  173
emotions and emotional reactivity,
  xxv, 55, 89, 91, 180
  brain and, 75
  eye contact and, 92
  Kagan experiment with babies, 56
  low reactivity, greater tolerance
    and, 58
  low reactivity, natural selection
    for, 66
  oxytocin and, 96, 234n25
  self-control and, 66
  theory of mind and, 5, 59
  understanding of false belief
    and, 57
empathy, 58
  beneficial contact and,
    249–50n71
  loss of, in face of threat to group,
    xxvi–xxvii, 111, 170
  oxytocin and reduced empathy
    toward rivals, 114, 232n5,
    234n25
Es Skhul Cave, Israel, 69
eugenics, 143–47, 194
evolution, xviii
  of brain size in hominins, 63
  cognitive, xxv, 55, 58–59, 62–63
  emotional reactivity and, 58
  friendliness and behavioral,
    morphological and cognitive
    change, 48
  genes and, 83–84
  of *Homo sapiens*, xviii–xix, 55, 65
  maximizing friendliness and
    cooperation, xvii, xxv–xxvi
  most successful species, xviii–xix
  natural selection and, xvii, xxvi
  patterns of growth and, 78
  survival of the fittest, xxv–xxvi

what it is, xviii
  See also human self-
    domestication hypothesis
eyes, 69, 72–77, 110
  black eyeballs as
    dehumanizing, 76
  brain neurons responsive to, 75
  of chimpanzees and bonobos,
    73, 73
  cooperative communication and,
    73–74, 76
  eye contact, 74, 76, 97,
    212–13n28, 218n34, 225n72
  human, colorful irises and white
    sclera, 72–77, 73, 78, 88, 97,
    225n72, 225n73
  "Kismet effect," 75–76
  Mickey Mouse and, 76
  of Mogwais (Gremlins), 77
  self-domestication hypothesis
    and, 76
  white sclera and cooperation, 76,
    225n72, 225n73

Facebook, 180
Fear Factor, The (Marsh), 234n25
Federalist Papers, 154
ferrets, 36
fish
  cavefish, 227–28n22
  cooperation and, 80
  physical and physiological traits
    related to friendliness and, 80
  self-domestication and
    friendliness of, 80, 227–28n22
  wrasse fish, 80
Fiske, Susan, 112–13
Flinders, Matthew, 154–55
foxes
  author's tests with, 27–31, 30
  Belyaev's experiment,
    domestication and friendly
    foxes, 19, 23–27, 24, 39,
    55–56, 81

cooperative communication and,
    27, 31, 55–56, 84
  genetic changes, 25–26
  gesture tests and, 30
  hormone changes, 67
  physiological and physical
    changes in friendly foxes
    (domestication syndrome),
    24–26, 49, 72, 84–85
  red and arctic, in urban areas,
    35–36
  retention of friendly juvenile
    behavior, 80, 81
  selection against aggression, 146
  selection for friendliness, 81
  serotonin in friendly foxes, 25, 85
  socialization of, 25, 81
  vocalizations of friendly foxes,
    81–82
Freedom Party, Austria, 245–46n33
Free Speech Movement, 179
friendliness
  aggression as less advantageous
    than, 42, 46–47, 80, 98
  author's tests with foxes,
    27–31, 30
  behavioral, morphological and
    cognitive evolution and, xvii,
    36, 48, 66–67
  Belyaev's experiment with foxes,
    19, 23–27, 24, 39, 55–56, 81
  in chickens, 26–27
  defined, xvii
  domestication and, xxv–xxvi,
    23–27, 24, 33–35, 51, 54,
    89–92, 203–4n8
  as evolutionary edge, xxvi, 67,
    99–100
  finger length and (2D:4D
    ratio), 51
  in fish, 80, 227–28n22
  flip side, cruelty to non-friends,
    xviii, xxvi–xxvii, 101, 102–21
  genetic changes and, 83–85

friendliness (*cont'd*):
  human, social networks and,
    66–67
  human evolution and, 65, 99
  human self-domestication
    and, 92
  in hunter-gatherer
    communities, 92
  juvenile-looking faces and, 68,
    69, 78
  members of a group and, 143
  in mice, 79–80
  natural selection and, xxv–xxvi,
    47, 48
  neural crest cells and, 85
  oxytocin and, 80
  physical and physiological traits
    related to, 67–77, 73, 78,
    80, 87
  serotonin and, 25, 51, 80, 85
  social development and, 82–83,
    94–99
  social tolerance and, 98
  strangers and, 93–94
  Williams syndrome, 26
  See also domestication; human
    self-domestication hypothesis
friendship, 171, 172, 195, 196

Galton, Francis, 144
Gariépy, Jean-Louis, 79
genes
  changes in, for friendliness,
    83–85
  changes in dogs, 17, 84
  changes in human, 84, 146–47
  evolution and, 83
  for development, 83–84
  librarian genes, 83–84, 85, 87,
    227–28n22
  Mendel's pea plants and, 83
  multitasking genes, 83, 84
genocide, 111, *115*, 115–16, 129,
    239–40n40

Germany, 129
  alt-right in, 245–46n33
  Asch's conformity experiment
    and, 132–33, *133*
  Eichmann trial, 134
  forced sterilization in, 145–46
  Goering on co-opting people,
    164
  hate speech laws, 180
  Holocaust and, 130, 132, 172,
    173
  rescuers who helped Jews,
    165–66
  traits of German society, 129
  Turks in, 131
Ghoshray, Saby, 142
Gingrich, Newt, xxviii–xxix
Goff, Philip, 136–37
  "Attitude, Inequality Mismatch,"
    130–31
  studies of prejudice, 141–42
gorillas, 38, 46, 126, 128, *128*,
    241n55
Greece, 245–46n33
groups
  agricultural communities,
    150–51, 243n3
  alt-right, 156–61
  Asch's conformity experiment
    and, 132–33, *133*
  beneficial contact to reduce
    intergroup conflict, 165–71
  cross-group friendships, 171
  cultural identity and, 166
  Degree of Structural Inequality,
    151
  identifying with, in children,
    110–12
  intolerance toward outsiders,
    158
  intragroup strangers and, 96–99
  model for large groups, 150–51
  modern societies as, 151
  oxytocin and, 96–97, 232n5

people's preference for their own group, 110–11, 143, 166
racial hierarchy and, 239–40n40
reciprocal humanization and, 170–71
rival groups or outsiders as a threat, 111, 120, 144, 158, 163, 164, 166, 167, 170, 181, 235n38
Robber's Cave experiments, 111
segregation vs. integration, 167, 248n58
social hierarchy and, 194, 195, 243n3
tendency prejudice, 111
tolerance and, 168–74, 249n70
"tyranny of the majority" and, 153
See also democracy, constitutional democracy; hunter-gatherer communities
Gruen, Margaret, 189–90
Gulf War, 121

Hadza people of Tanzania, 92–93
Hamilton, Alexander, 154
hamsters, 109
hands
androgens and, 51, 69–70, 88
2D:4D and, 51, 70
Harari, Yuval, 100
Harris, Lasana, 112–13
Hemmer, Helmut, 22
Henrich, Joseph, 95–96
Herculano-Houzel, Suzana, 61–62
Heyer, Heather, 178–79
Hill, Kim, 98
Hill and Knowlton PR firm, 121
Hodson, Gordon, 169–70, 190
Hoffman, Kelly, 140
Homo erectus, xx, 72, 119
stone tools, Achulean hand ax, xx, xxi

Homo sapiens (humans)
advances by, Upper Paleolithic period, xxiii–xxiv
aggression against threats, 110, 232n5
amygdala of, 56, 57
behavioral modernity and, xxiv, 64, 67, 222n353
brain and brain size, xix, 63, 64, 66, 70, 88, 90–91, 99, 119
Churchill's analysis of Pheistocene skulls, 69, 69n
cognitive evolution in, 55, 59, 203–4n8
comparison of human, bonobo, and chimpanzee babies, 88–89
cooperative communication and, 56, 57, 58, 66, 69, 78, 91–92
culture, settlements, and society, xxiii–xxiv, xxvi, 63, 67
dehumanizing others and, xxvi–xxvii, 110–14, 117, 117–20, 118
development, patterns of growth, 88–89
emotional reactivity in, 56, 57, 58, 66
evolution of, xviii–xix, 55, 65
eyes, colorful irises and white sclera, 72–77, 73, 78, 88, 97, 225n72, 225n73
facial features in, 67, 68–69
fossil record and, 68, 72
friendliness, as evolutionary edge, xxvi, 67, 99–100
friendliness, physical and physiological traits, 67–77, 73, 78
friendship and survival, 196
gap between first appearance and population, cultural explosion, xix
genetic changes in, 84, 146–47

Homo sapiens (humans) (*cont'd*):
  group identity and social bonds,
    95, 110
  hands of, 51, 70
  human-animal bonds and,
    191–96
  intelligence and, 63–64, 88,
    203–4n8
  judging based on physical traits,
    189
  love of dogs and, 191–96,
    212–13n28
  neural crest cells and, 87
  other human species and, xix,
    xxi, 63–64, 77
  pigmentation of, 72–77, 87
  pointing, as fundamental skill,
    3–4, 7
  population density and, 37, 65,
    69, 182, 222n35
  secret to survival of, xxiv–xxv
  self-control and, 63, 66, 90, 99,
    100
  self-domestication and, 55–77,
    87, 89–92, 99, 109, 222n35
  similarity to dogs, 15, 16–17, 32
  skull shape of, 71, 72, 88, 90,
    119
  social networks and, xxiii, 63–64,
    65, 92, 94–98
  strangers and intragroup
    strangers, 93–100, 110
  success of, xxi–xxii
  technology, tools, weapons,
    ornaments, xix, xxi–xxii,
    64–65, 67, 69, 98, 147
  theory of mind in, 59
  as ultracultural species, 66–67
  as urban species, 37, 182–85
  years of existence, xix
  See also human self-
    domestication hypothesis
Howell, F. Clark, 116

human. See *Homo sapiens*
human cruelty
  Ascent of Man Scale and,
    116–17, *117*
  Banyamulenge tribe massacre,
    102–8
  dehumanization and, 110–14,
    117, 117–20, *118*, 134–37
  explanations for atrocities,
    132–37
  fearful, aggressive behavior
    toward strangers, 114–20
  genocide, 111, *115*, 115–16
  toward animals, as predictive
    toward people, 190
human genome, 147
human nature, xvi, 145, 149, 154,
    165, 185, 201
human self-domestication
    hypothesis, xxx
  beneficial contact and, 170,
    249–50n71
  cognitive advances and, 58–59
  contact between groups to
    remove dehumanization and,
    170
  dehumanization, tendency
    prejudice and, 111, 161,
    162–63, 176, 180
  dehumanization and acts of
    violence, 180, 181
  friendliness toward intragroup
    strangers and, 97
  "mutual animosities" as key
    feature, 153
  paradox of human nature,
    143–44, 165
  peaceful protests and, 176
  predictions of, 66, 76, 161, 176,
    180
  reciprocal dehumanization, 142,
    170
  tendency prejudice and, 111

thriving as a result of
friendliness, 67
ultracultural species and, 66
what it is, xxx, 65–66
white sclera and, 76
Hungary, 132
Hunt, James, 127
hunter-gatherer communities,
243n3
aggression toward strangers
and, 93
Baka pygmies, 122
dogs as family and, 193, 195
as egalitarian, 150, 193
friendliness in, 92
infanticide among, 144
lethal violence by, 115
Martu Aboriginals, 191–93
social networks in, 92–93
Hurston, Zora Neale, 167

Ignacio, Natalie, 28–31, 30
intelligence. See cognition
intragroup stranger, 94–99, 110
Inuits, 65, 97, 144
Israel and Israelis, 119, 120, 142,
180

Jacobs, Jane, 183
James, LeBron, 138
Japan, 129, 132
Jardina, Ashley, 139–40, 142, 160,
163
Jay, John, 154
Jobbik Party, Hungary, 245–46n33

Kagan, Jerome, 56
Kayş, Roland, 214n43
Kelly, Sean, xxix
Key deer, 36
Khrushchev, Nikita, 172–74
kindness, xvii, 46, 94, 99, 121, 143,
157, 165, 189–91

Kindness Clubs, 188–89
King, Martin Luther, Jr., 176
King Kong (film), 127–28, 128, 138
Kteily, Nour, 116–17, 118, 119,
120, 157–58
March of Progress scale and,
116, 116, 138–39, 159
Kukekova, Anna, 26
Kurzweil, Ray, 148

language, xxiii, xxv, 4, 6, 52, 62, 99,
123
cultural identity and, 166
dehumanizing, xxviii–xxix, 137,
180–81
low reactivity in children and, 57
Leach, Jim, xxvii
Leibovitz, Annie, 138
Le Pen, Marine, 245–46n33
Linnaeus, Carolus, 126
Lola ya Bonobo, Republic of
Congo, 42, 43–44, 49–51, 82
Lyall, Jason, 176

MacLean, Evan, 60
Madison, James, 153, 154
Malcolm X, 176
Mandela, Nelson, 160
March of Progress scale, 116,
116–17, 117, 138–39, 158,
159
Marsh, Abigail, The Fear Factor,
234n25
marshmallow test, 59–61, 90
Martu Aboriginals, 191–93
Dreamtime, 192
McCain, John, xxix
McCarthy, Andrew, 132
McLoughlin, Niam, 95
mice, 79–80
Mickey Mouse, 76
microbiome, xvii–xviii
Miklosi, Adam, 212–13n28

Milgram, Stanley, experiments on
    obedience to authority,
    133–34, 136
Millennium Project, 148
mitochondria, xvii
Mori, Masahiro, 125
Muslims, 117–18, 120, 142–43,
    159, 235n38, 245–46n33
Mutombo, Dikembe, 138

natural selection
    Darwin and, xvii, 20–21
    friendliness and, xxv–xxvi, 47,
        48, 65–66, 68, 76, 81,
        227–28n22
    self-domestication and, xxvi,
        37, 38
    shaping our cultural cognition, 58
Neanderthals, xx–xxi, xxii, 99
    extinction and, 63
    eyes of, 77
    Homo sapiens vs., 63, 99
    size of groups of, 182
    skull of, 72, 90, 119
    2D:4D and, 70
Nelson, Emma, 69
Netherlands, 245–46n33
neural crest cells, 84–87, 86,
        228n26
    adrenal glands and, 85
    brain development and, 86
    neurocristopathy and, 87
    selection for friendliness and,
        85, 86
Norquist, Grover, 155
Northwest Coast Native
    Americans, 243n3
Nyugen, Mai, 184

Obama, Barack, 138, 241n55
Obama, Malia, 241n55
Obama, Michelle, 241n55
Okanoya, Kazuo, 36–37

Oliner, Samuel and Pearl, 166
Omar, Ilhan, xxvii
O'Neill, Tip, xxvii
orangutans, 51
Oreo (author's dog), 5, 8, 9, 196,
    197
    pointing and, 8–10, 11, 11–14
Origin of Species (Darwin),
    xvi–xvii, 20
Ornstein, Norman, xxviii
oxytocin
    aggression and, 109–10, 113,
        232n5
    in babies and parents, 73–74
    brain development and, 86
    extended juvenile development
        and, 80
    eye contact and, 76, 97
    intragroup stranger and, 96–97
    mother-child bonding and, 97
    negative treatment of outsiders
        and, 112–14, 232n5, 234n25
    serotonin and, 96, 110, 231n54
    social bonds and, 109

Paine, Thomas, 153, 156
Palestinians, 119, 120, 142
parsimony, principle of, 10
People to People, 172, 174
Petry, Frauke, 245–46n33
Pinker, Steven, 100
Pitynski, Andrzej, 165
Plato, 144
    Republic, 156
pointing
    animal experiments, 10–14
    author's gesture tests with foxes,
        28–31, 30
    author's gesture tests with Oreo,
        8–14, 11
    chimpanzees and, 6, 14–16
    cooperative communication and,
        6–7

as developmental skill in babies, children, 3–4, 6–7, 14
dogs' understanding of, 13–14, 31–32, 212–13n28
experiment with babies, 14
polar bears, 109–10
prejudice, 132, 136, 140–42, 160, 238n27
    Allport's theory of prejudice, 132–33
    beneficial contact to stop it, 165–71, 248n58, 249n70
    factors driving, 189–90
    racism and, 129–30, 136–42, 137, 239–40n40, 240n42, 241n55
    tactics, ineffective to reduce, 160–61
    tendency prejudice, 111
    See also dehumanization
pygmies, 122–23

racism. *See* prejudice
rats, 109
Reagan, Ronald, xxvii
reciprocal dehumanization, 142–43, 170
    blacks and whites, 142
    nonviolent vs. violent protests, 176
    Palestinians and Israelis, 142
reciprocal humanization, 170–71
Republic (Plato), 156
rhesus macaque, 63
Right Wing Authoritarianism (RWA), 156–61, 158, 162
    beneficial contact to reduce intergroup conflict and, 170, 170
Robber's Cave experiments, 111
Roma (Gypsies), 118–19, 159
Romania, 132
Rostenkowski, Dan, xxvii

Russia, 129
    Great Terror of 1937–38, 18
    World War II atrocities, 132
Rwanda, 104–5, 186

salamanders, 78, 84
Sartre, Jean-Paul, 176
Scottsboro Boys, 128, 128–29
self-control, 59–63
    in animals, 60
    brain's executive center and, 59
    brain size, neuron density, and, 61–62
    human character traits and, 60
    in humans, 63, 66, 90–91, 99, 100
    marshmallow test and, 59–61
    studying variations in, 59
self-domestication, xxx, 55–77, 97
    aggression toward strangers and, 109
    androgens and, 96
    of bononos, xxvi, 48–49
    of cavefish, 227–28n22
    of coyotes, 35
    dehumanization and, 111–12
    development of juveniles and, 78–92
    of dogs, xxvi, 33–35
    evolution of genes and, 83–84
    friendliness and, xxvi, 33–35, 54, 97, 100–101, 101
    of humans, xxvi, 55–77, 87, 89–92, 96, 97, 182, 222n35
    increasing population density of humans and, 37, 65, 182, 222n35
    intragroup stranger and, 97
    morphology, physiology, and cognition changes, 66
    natural selection and, xxv–xxvi, 47, 48, 65–66, 68, 76, 81, 227–28n22

self-domestication (cont'd):
  urban environments and, 35–37,
    182–85
  what it is, xxv–xxvi
  what it will predict, 66
  white sclera and, 76, 77, 97
  of wolves, 33–35, 48
  See also human self-
    domestication hypothesis
serotonin
  aggression and other behavior
    effects, 25, 51, 71, 110,
    227–28n22
  in bonobos, 51
  brain development and, 86,
    227–28n22
  brain size and, 70, 110
  deficiency and criminal
    behaviors, 71
  Ecstasy (drug) and, 71
  extended juvenile development
    and, 80
  in friendly foxes, 25
  oxytocin and, 96, 110, 231n54
  skull shape and, 71, 72, 88
  SSRIs and, 71
Sherif, Muzafer, Robber's Cave
  experiments, 111
Sherwood, Chet, 51
slavery, 126, 127
Social Dominance Orientation
  (SDO), 156–61, 158, 159,
    162
  beneficial contact to reduce
    intergroup conflict and, 170,
    170
  dog SDO survey, 194–95
social networks
  Australian vs. Tasmanian
    Aboriginals, 64–65
  feedback loop of, 65
  group identity and, 94–99
  in hunter-gatherer communities,
    92–93
  intragroup stranger and, 94
  Inuits, isolated, and, 65
  population density and, 65
  rapid expansion, 63–64
  technology development and,
    64–65
  tolerance and, 94
social skills, 78–79, 82–83
  cooperation, 92–93, 100
  fetal brain growth and, 90
  human babies vs. chimpanzee
    and bonono infants, 89
Sorcerer's Apprentice, The (film), 76
Spencer, Richard, 176
SSRIs (selective serotonin reuptake
  inhibitors), 71
Stables, Gordon, 193
Stalin, Joseph, 18
Steamboat Willy (film), 76
Stenner, Karen, 160
Stowe, Harriet Beecher, Uncle
  Tom's Cabin, 169
strangers, 92–98
  aggressive, fearful behavior
    toward, 109–12, 114–20,
    232n5
  Ascent of Man Scale and,
    116–17, 117, 118
  bonobos and, 50, 93
  chimpanzees and, 50, 93
  dehumanization of, 110–12, 117,
    117–20, 118
  the dehumanizing brain and,
    112–14
  eye contact and, 97
  group identity and, 95
  human capacity for friendliness,
    increase in, 100–101, 101
  human cultural innovation
    and, 97
  intragroup stranger, 94–99
  mice and, 79–80
  oxytocin and, 96–97, 99–100,
    234n25

psychopaths vs. altruists and response to, 234n25
rival groups, persecution of, 102–8
as a threat, 120, 144
Sullivan, Andrew, 156
survival of the fittest
    alt-right's belief in caricature of, 156–57
    cooperative communication and, xxv–xxvi
    defined by Darwin and modern biologists, xvi
    misinterpreted as strong and ruthless, xvi–xvii
    Stalin's misconception, 18
Sweden Democrats, 245–46n33

Tasmanian Aboriginals, 64, 97
Tchernov, Eitan, 22
technology
    cooperative communication and, 65–66
    friendliness in the Pleistocene and, 65, 69
    Homo sapiens and, xix, xxi–xxii, 64–65
    human self-domestication hypothesis and, 65
    as ineffective for solving social problems, 149
    as problem, 148–49
    rate of accelerating change, 147–48
    social networks and development of, 64–65
temperament, theory of mind and, 58
tendency prejudice, 111
termites, 79, 84
testosterone
    aggression, competition and, 50–51, 67–68, 70
    in bonobos, 82

hand features or 2D:4D and, 51, 69–70, 88
    masculinized faces and, 67, 68, 88
    serotonin, oxytocin and, 96, 231n54
Theofanopoulou, Constantina, 87
theory of mind
    in animals, 6, 51–52
    in children, 5–6, 56, 95
    compassion and, 101
    cooperative communication and, 4, 27
    false belief, 57
    human brain network for, 57, 75, 90, 110, 112–14, 119
    in humans, 59, 63
    negative treatment of outsiders and, 112–14
    pointing and, 3–4, 6
    temperament and, 58
    what it is, 4–5
Tibetan mastiff, 17
Tomasello, Mike, 5–8, 10, 39, 59, 64
    theory of mind study, 5–6
Topál, József, 36
Trump, Donald, 174, 181, 245–46n33
Trump, Donald, Jr., xxvii
Trut, Lyudmila, 19, 24, 26, 27
Tsimane of Bolivia, 94

uncanny valley, 125, 125–26
Uncle Tom's Cabin (Stowe), 169
United Nations Refugee Agency (UNHCR), 106–7, 108
University of California, Berkeley, 179
urban environments
    animals in, 35–36
    best architecture for, 182–83
    "hostile architecture," 184
    humans as urban species, 37, 182–85

urban environments (*cont'd*):
  negative aspects for social
    contact, 183–84
  promoting tolerance in, 184–85
  racial segregation in, 183
U.S. Congress
  approval rates, 244n25
  bipartisan friendliness and,
    xxvii–xxviii, xxx, 207–8n48
  Hastert rule, 164
  members' place of residence and,
    207–8n48
  polarization of, xxvii–xxx,
    164–65, 207–8n48
  pork barrel projects, xxix–xxx
  staffers, 207–8n48
U.S. Constitution, 154, 155
  First Amendment, 179–80
U.S. military
  beneficial contact in, 168
  desegregation of, 168
U.S. presidency, 244n25
U.S. Supreme Court, 244n25

*Variations of Plants and Animals
  Under Domestication, The*
  (Darwin), 20–21

Walker, Kara, 44
Welch, Kelly, 239–40n40
Wellman, Henry, 56–57
white-rumped munia, 37
white-tailed deer, 36

wildebeest, 91
Wilders, Geert, 245–46n33
Wilkinson, Adam, 85, 86
Williams syndrome, 26
wolves, 23, 228n26
  barking in, 81
  dogs evolving from, 17
  extinction threat, 35
  human cooperative
    communicative intentions
    and, 32, 212–13n28
  human feces and, 34
  Miklosi's comparison of wolf and
    dog pups, 212–13n28
  self-domestication of, 33–35,
    48, 99
Women's March, 174, *175*,
  175–76
Woodhull Martin, Victoria, 144
World War II
  Holocaust and, 130, 132, 172,
    173
  Japanese atrocities, 132
  rescuers who helped Jews,
    165–66
Wrangham, Richard, 27, 30, 39,
  41, 58, 85, 86, 222n35
wrasse fish, 80, 84, 228n26

Yerkes research center, 39–42
Yiannopoulos, Milo, 179

Zhou, Wen, 194

About the Authors

BRIAN HARE is a core member of the Center of Cognitive Neuroscience and a professor in evolutionary anthropology, psychology, and neuroscience at Duke University. He received his Ph.D. from Harvard University, founded the Hominoid Psychology Research Group while at the Max Planck Institute for Evolutionary Anthropology in Germany, and subsequently created the Duke Canine Cognition Center when arriving at Duke. He has co-authored three books and published over a hundred scientific papers, including in *Science, Nature,* and *PNAS*. His research on dozens of different animal species, including dogs, wolves, bonobos, chimpanzees, and humans, has taken him everywhere from Siberia to the Congo Basin. His work has received consistent national and international attention. In 2007, *Smithsonian Magazine* named Hare one of the world's top 35 scientists under 36. He and his research have been featured on *60 Minutes, NOVA,* and *Nature,* as well as the series *Is Your Dog a Genius?,* which he hosted for *National Geographic Wild*. Most recently, in 2019, he participated in Steven Spielberg's documentary series *Why We Hate,* which was released worldwide on Discovery Channel.

VANESSA WOODS is a research scientist in evolutionary anthropology at Duke University and an award-winning writer and journalist. Woods has written eight nonfiction books for both adults and children. In 2010, her book *Bonobo Handshake: A Memoir of Love and Adventure in Congo* won the Thomas Lowell Award for nonfiction, and her children's book *It's True, Space Turns You into Spaghetti* was named an Acclaimed Book by the Royal Society in 2007. Her books have been translated into twelve languages. Woods received the Australiasian Science Award for journalism in 2004. Her articles have been published in both national and international media, including *The Wall Street Journal, National Geographic,* and *The New York Times*.

Hare and Woods are married and live with their family and dog, Congo, in North Carolina. Their first book together, *The Genius of Dogs,* was a *New York Times* bestseller.